The Sky is Falling

Shane Frye

Published by Shane Frye, 2026.

THE SKY IS FALLING

First edition. February 2, 2026.

ISBN: 979-8233293627

Written by Shane Frye.

Table of Contents

The Sky is Falling: Omens of the Atmosphere

By Shane Frye

Introduction

Humanity has always looked upward for answers. Long before satellites tracked storm systems or meteorologists predicted rainfall percentages, our ancestors read the sky with an intimacy born of necessity. They observed patterns in clouds, noted shifts in wind direction, measured the quality of light at dawn and dusk. Their survival depended on interpreting atmospheric signals correctly. A misjudged sky could mean death at sea, failed harvests, or settlements destroyed by sudden floods.

What ancient peoples called omens, we now understand through atmospheric science. Yet the fundamental observations remain unchanged. A green tinted sky still precedes tornadic activity. A halo around the moon continues to signal approaching precipitation. The blood red moon that terrified medieval populations results from the same particulate matter that concerns modern climatologists. These atmospheric phenomena bridged the gap between superstition and science, between prophecy and prediction.

This book examines fifteen atmospheric signs that have shaped human history, influenced military campaigns, triggered religious movements, and altered the course of civilizations. Each chapter explores both the meteorological mechanisms behind these phenomena and their historical impact. We will trace how leaders interpreted celestial events, how populations responded to unusual skies, and how atmospheric conditions correlated with societal upheaval.

The sky speaks in a language older than writing. Cirrus clouds telegraph weather systems days in advance. The scent of ozone announces electrical storms. Dust devils reveal soil conditions and wind patterns. These signs require no instruments to detect, only attention and understanding. Our ancestors possessed this literacy by necessity. We have largely lost it through technological dependence.

Yet these omens remain relevant. Climate change intensifies atmospheric extremes. Solar storms threaten electrical infrastructure. Atmospheric rivers produce floods of biblical proportions. The same signs our ancestors watched now carry different urgency. Understanding them connects us to both deep history and immediate survival.

Modern meteorology confirms what traditional knowledge long maintained: the atmosphere broadcasts information continuously. Pressure systems, humidity levels, particulate content, electromagnetic activity, all leave visible traces. Learning to read these signs doesn't require rejecting science for superstition. Rather, it means recognizing that empirical observation preceded theoretical understanding by millennia.

The chapters ahead examine atmospheric phenomena from two perspectives simultaneously. First, the scientific explanation: what physical processes create these visible signs. Second, the historical context: how populations interpreted and responded to them. This dual approach reveals how atmospheric conditions influenced human behavior across cultures and centuries.

Some phenomena discussed here remain imperfectly understood. Scientists debate whether auroras produce audible sounds. The exact chemistry of the green flash continues to generate research papers. Electrophonic hearing challenges conventional physics. These mysteries remind us that the atmosphere still guards secrets, that observation and wonder need not exclude each other.

This book assumes no prior scientific knowledge. Complex atmospheric processes are explained in accessible language. Historical accounts come from primary sources where possible. The goal is literacy in sky reading, understanding both mechanism and meaning.

The atmosphere connects all life on Earth. Weather patterns cross political boundaries. Solar radiation affects every organism. Atmospheric composition determines habitability. Learning to read the sky means developing fluency in a universal language, one spoken before humans existed and which will continue long after we're gone.

Each chapter can be read independently, though together they form a comprehensive survey of atmospheric omens. Some focus on specific weather phenomena. Others examine celestial events visible through our atmosphere. All share the common thread of observable signs carrying actionable information.

The sky is falling. It always has been. Water molecules precipitate. Cosmic rays penetrate. Electromagnetic radiation cascades. Particles circulate. These processes created the conditions for life and continue sustaining it. Understanding atmospheric omens means recognizing our perpetual immersion in dynamic chemical and physical systems.

Our ancestors knew this intuitively. They watched skies with attention we reserve for screens. They noticed subtle shifts we ignore. They developed predictive systems that worked, even without understanding why. This book recovers that lost knowledge while adding modern scientific context.

The following chapters explore specific atmospheric phenomena in depth. Each examines historical interpretation alongside contemporary explanation. The result is a bridge between ancient sky lore and current atmospheric science, between omen and observation, between prophecy and prediction.

Look up. The sky speaks constantly. Learning its language opens perspectives spanning millennia, connecting human experience across vast temporal and cultural distances. These omens shaped civilizations. They still shape our world. Understanding them enriches both historical comprehension and present awareness.

The atmosphere remembers everything. Ice cores preserve ancient air. Tree rings record drought cycles. Sediment layers archive volcanic eruptions. The sky writes history in multiple mediums simultaneously. Reading these records requires knowing what to look for and where to find it.

This book serves as introduction and guide. It cannot replace direct observation. Understanding atmospheric omens demands time spent watching skies, noting patterns, correlating signs with outcomes. Theoretical knowledge supports but never substitutes for empirical experience.

Atmospheric phenomena are inexhaustible. Every culture developed localized sky reading traditions. Every region exhibits unique meteorological patterns. These chapters focus on widely observable signs with documented historical impact, but they scratch the surface of atmospheric literacy.

Read, then observe. Test these explanations against lived experience. The sky remains the most accessible natural phenomenon, visible to everyone, constantly present. Its omens continue broadcasting. We need only recover the ability to receive them.

Chapter One: The Green Sky

The green sky terrifies with good reason. When the atmosphere takes on that distinctive sickly chartreuse hue, conditions are optimal for tornadic development. This coloration has sent populations fleeing to shelter for centuries, though the scientific explanation remained elusive until recent decades. The green sky represents one of the most reliable atmospheric omens, a visual warning system nature provides free of charge.

The phenomenon occurs during severe thunderstorms, specifically those with rotating updrafts called supercells. These storms produce the most violent tornadoes, hail of extraordinary size, and damaging straight line winds. The green color appears most prominently in late afternoon or early evening, when sunlight angles through the storm at specific geometries. Not every severe storm produces green skies, and not every green sky spawns tornadoes, but the correlation is strong enough to warrant immediate protective action.

Understanding why skies turn green requires examining light interaction with water and ice particles suspended in storm clouds. Thunderstorms contain enormous quantities of liquid water and frozen precipitation. A typical supercell holds millions of tons of water in various states. This water content creates a massive optical filter through which sunlight must pass.

White light from the sun contains all visible wavelengths. When this light enters a thunderstorm, water droplets and ice crystals scatter and absorb different wavelengths selectively. Red wavelengths, being longer, penetrate deeper into the storm than shorter blue wavelengths. The storm's internal structure bends, reflects, and filters this light repeatedly before it exits and reaches observers on the ground.

The green coloration results from specific combinations of factors. First, the storm must contain exceptionally large concentrations of water. Second, significant hail must be present, particularly large hailstones suspended in powerful updrafts. Third, the sun must be positioned at low angles, typically late in the day. Fourth, the observer must be positioned where light exits the storm after passing through its core.

Research by meteorologists in the 1990s used spectrophotometers to measure the actual wavelengths present in green storm light. They confirmed that the predominant wavelength matched cyan to green portions of the visible spectrum. The measurements revealed that red light from the low sun passed through the storm, mixing with blue light scattered by water droplets and ice particles. This combination produces the characteristic green hue.

The presence of hail plays a crucial role. Large hailstones, some exceeding softball size, remain suspended in supercell updrafts reaching speeds over 100 miles per hour. These ice spheres scatter blue light efficiently while allowing red light penetration. The specific size distribution of hailstones affects the exact shade of green observed. Storms producing baseball sized or larger hail often generate the most vivid green coloration.

Historical records document green skies preceding catastrophic tornadoes. The Tri State Tornado of 1925, which killed 695 people across Missouri, Illinois, and Indiana, was preceded by reports of green black skies. Survivors described the color as unlike anything previously witnessed, a sickly luminescence that induced visceral dread. Similar accounts accompany nearly every major tornado outbreak throughout recorded history.

Native American tribes of the Great Plains recognized green skies as harbingers of destruction. Oral traditions passed down through generations emphasized seeking shelter immediately upon observing this coloration. These communities lived in close relationship with atmospheric conditions, their survival dependent on reading weather signs accurately. The green sky held such significance that some tribes incorporated it into spiritual practices, viewing it as a physical manifestation of dangerous spirits.

European settlers initially dismissed indigenous weather lore as superstition. Lacking cultural context for North American tornado climatology, early colonists often ignored warning signs that native populations understood implicitly. Death tolls from tornadoes ran proportionally higher among settler communities until they developed their own observational traditions, often learning from indigenous neighbors.

The deadliest tornado in United States history, the Natchez Mississippi tornado of 1840, arrived under green skies so dark that witnesses reported being unable to see their hands at arm's length. The tornado killed 317 people directly, with the death toll rising to 425 when accounting for victims who died aboard destroyed river boats. Survivors' accounts emphasize the preternatural darkness and green illumination that preceded the tornado's arrival.

Not every green sky produces a tornado. Storms can exhibit green coloration without developing rotation. However, the presence of green light indicates that conditions favorable for severe weather exist. The storm contains massive water content, powerful updrafts, and likely significant hail. Even without a tornado, such storms pose serious threats through large hail, damaging winds, and dangerous lightning.

Modern tornado warnings incorporate visual observations alongside radar data. Trained storm spotters specifically watch for green sky coloration as one indicator among many. The National Weather Service encourages the public to seek shelter not just when sirens sound but when skies take on unusual colors, particularly green or yellow green hues.

The psychological impact of green skies cannot be overstated. Something fundamental in human perception responds to this coloration with alarm. The color exists outside normal experience, triggering instinctive wariness. This response serves survival, prompting protective action before conscious analysis occurs. Populations that paid attention to this instinct survived to pass along their observations.

Photographic documentation of green skies proves challenging. Camera sensors and human eyes process color differently. Many photographs of green sky storms appear gray or merely dark to viewers, failing to capture the distinctive hue visible to observers present during the event. This technical limitation means written descriptions provide better historical records than most photographic evidence.

The greenness varies in intensity and shade. Some storms produce pale mint coloration. Others generate deep emerald or olive tones. The most ominous appear as bruised yellow green, a color that seems to exist outside natural spectrums. These variations reflect differences in storm structure, hail content, water concentration, and light angles. Experienced storm observers distinguish between shades, using color intensity as one factor in threat assessment.

Tornadoes can form without green skies, particularly at night when color perception becomes impossible. Nocturnal tornadoes kill disproportionately because visual warning signs disappear. The green sky omen functions primarily during daylight hours, providing advance notice that supplements but never replaces other indicators like wall clouds, rapidly rotating cloud bases, or the characteristic roar of approaching tornadoes.

The duration of green coloration varies. Some storms maintain the color for hours as they move across terrain. Others exhibit it briefly, sometimes for only minutes before tornado genesis. Rapid intensification of green hue often correlates with strengthening rotation and increased tornado probability. Observers who monitor color changes gain additional insight into storm evolution.

Children often notice green skies before adults, their attention less filtered by routine and expectation. Numerous accounts describe children alerting parents to the unusual sky color, prompting families to seek shelter minutes before tornadoes struck. This pattern suggests that cultivating atmospheric awareness from young ages enhances community safety.

The green sky phenomenon extends beyond tornado production. Severe straight line wind events called derechos sometimes occur under green illuminated skies. These windstorms produce tornado level damage across much wider paths, devastating forests and urban areas alike. The green coloration signals severe convective activity regardless of specific storm type.

Lightning interacts with green skies to create memorable visual effects. The green backdrop makes lightning appear brilliant white or blue white, enhancing contrast dramatically. This combination of green sky and intense lightning further amplifies the ominous atmosphere preceding severe weather. Observers report that time seems to slow during these conditions, perception heightening in response to environmental threat cues.

Regional variations in green sky appearance reflect local atmospheric characteristics. Great Plains storms exhibit different green tones than those in the Deep South or Midwest. Humidity levels, local pollution, and typical storm structures all influence specific coloration. Experienced observers in each region develop calibrated responses to their area's particular green sky manifestations.

The green sky omen represents intersection of physics, perception, and survival. Light wavelengths, water content, ice crystal properties, and human visual systems combine to create an effective early warning system. No technology required, no special training necessary. Simple observation of color provides actionable information with life or death implications.

Modern Doppler radar detects rotation within storms with greater precision than human observation alone. However, radar cannot replace visual monitoring. Technical failures occur. Power outages disable warning systems. Human observation provides redundancy and fills gaps in technological coverage. The green sky omen functioned for centuries before radar existed and will continue functioning if technological systems fail.

Education about atmospheric omens improves community resilience. Populations taught to recognize green skies as danger signals respond appropriately even when official warnings fail to reach them. This knowledge distributes safety information beyond technological channels, empowering individuals to make protective decisions based on direct observation.

The next time skies take on that distinctive green cast, understand what you're seeing: light filtered through millions of tons of suspended water and ice, a visible manifestation of atmospheric violence. The color itself will not harm you, but it signals conditions that absolutely can. Seek substantial shelter immediately. Bring others with you. Wait until the storm passes and skies return to normal coloration before emerging.

The green sky remains one of nature's clearest warnings, a gift of physics and chemistry that requires only attention to receive. Our ancestors heeded this signal because their lives depended on it. Ours still do.

Chapter Two: Halo of the Moon

The luminous ring encircling the moon has inspired poetry, prophecy, and practical weather forecasting for millennia. This optical phenomenon, visible to anyone who looks upward on the right kind of night, signals approaching precipitation with remarkable reliability. Ancient mariners, farmers, and travelers learned to read this celestial sign, understanding that the halo's appearance meant rain or snow would arrive within roughly two days. Modern atmospheric science confirms what traditional knowledge long maintained: the moon's halo results from ice crystals in high altitude clouds, the leading edge of weather systems bringing moisture.

The halo appears as a perfect circle of light surrounding the moon, typically spanning twenty-two degrees of arc when measured from the observer's perspective. This specific angular diameter results from the geometric properties of ice crystals and light refraction. The ring glows softly, usually white or pale, though occasionally displaying subtle color separation with reddish inner edges and bluish outer boundaries. On clear nights between storm systems, this ethereal crown transforms the moon into something otherworldly, beautiful and ominous simultaneously.

Ice crystals in cirrus clouds create the halo effect. These clouds form at altitudes between 20,000 and 40,000 feet, where temperatures remain well below freezing year round. Water vapor at these heights crystallizes into hexagonal ice structures, millions upon millions of six sided prisms floating in the upper atmosphere. When moonlight passes through these crystals, refraction bends the light at specific angles determined by ice crystal geometry.

The hexagonal structure of ice crystals is not arbitrary but follows from water molecule bonding patterns. Each H2O molecule bonds with four neighbors in a tetrahedral arrangement. When water freezes, these bonds lock into repeating hexagonal lattices. The resulting crystals naturally form six sided columns and plates. These geometric properties remain constant across all ice crystals, ensuring consistent optical effects regardless of where or when the crystals form.

Light entering an ice crystal bends as it crosses from air into ice, then bends again exiting back into air. The minimum angle at which light can bend through a randomly oriented hexagonal crystal is approximately twenty-two degrees. Light bent at smaller angles gets trapped within crystals through internal reflection. Light bent at larger angles disperses too widely. The concentration of refracted light at twenty-two degrees creates the visible halo ring.

Millions of ice crystals must be present for a halo to appear bright enough for human perception. Individual crystals produce tiny points of refracted light, but when countless crystals align across the sky, their collective effect becomes visible. The crystals need not be perfectly aligned; random orientation still produces halos because the twenty-two degree minimum refraction angle applies regardless of individual crystal positioning.

Cirrus clouds typically precede warm fronts by one to two days. These high altitude ice clouds stream ahead of approaching low pressure systems, announcing their arrival long before lower, thicker clouds appear. The sequence is predictable: cirrus clouds first, followed by cirrostratus, then altostratus, and finally the nimbostratus or cumulonimbus clouds that produce actual precipitation. The moon halo appears during the cirrus or early cirrostratus stage, providing advance warning.

The reliability of the "ring around the moon, rain by noon" saying varies by location and season. In temperate zones during fall, winter, and spring, the correlation between moon halos and subsequent precipitation exceeds seventy percent. Summer patterns prove less reliable as isolated thunderstorms develop without the organized frontal systems that generate predictable cloud sequences. In tropical regions, different meteorological patterns reduce halo accuracy as a precipitation predictor.

Historical records document reliance on moon halos for agricultural planning. Medieval European farmers watched for halos to time planting and harvesting activities. A halo appearing during critical agricultural periods could prompt rushed efforts to bring in crops before rain arrived. Maritime cultures placed even greater emphasis on this sign, as unexpected storms at sea carried far graver consequences than wet harvests on land.

Ancient Greek and Roman writers referenced moon halos in weather prognostication texts. Aristotle's Meteorologica discusses lunar crowns and their association with coming rain. Pliny the Elder's Natural History catalogs various celestial signs including moon halos, noting their reliability for weather prediction. These classical sources indicate that sophisticated sky reading traditions existed throughout the Mediterranean world.

Chinese astronomical records spanning thousands of years include detailed moon halo observations. Ancient Chinese astronomers classified halos by color, brightness, and angular diameter, developing complex interpretive systems linking halo characteristics to weather outcomes. These records reveal that careful empirical observation occurred across multiple independent cultures, all reaching similar conclusions about halo significance.

Native American traditions across North America recognized moon halos as rain signs. Tribal groups from diverse regions developed similar interpretations despite having no contact with each other or with Old World cultures. This independent convergence on identical meanings suggests the phenomenon's reliability transcends cultural interpretation, reflecting genuine atmospheric conditions visible to any observer.

The physics of halo formation means that their appearance requires specific atmospheric conditions. The upper atmosphere must be cold enough for ice crystal formation but not so cold that moisture content drops too low. Sufficient water vapor must be present to generate extensive cirrus clouds. These conditions correlate strongly with approaching weather systems, particularly warm fronts where moist air gradually overrides cooler air masses.

Color separation in moon halos follows the same principles as rainbow formation. Different wavelengths of light refract at slightly different angles through ice crystals. Red light, having the longest wavelength, bends least and appears on the halo's inner edge. Blue light bends more, appearing toward the outer edge. However, these colors usually appear faint because moonlight is much dimmer than sunlight, providing less intense illumination for ice crystal refraction.

Sun halos occur through identical mechanisms, ice crystals refracting sunlight instead of moonlight. The much greater intensity of solar illumination makes sun halos dramatically brighter and more colorful than their lunar counterparts. However, looking directly at sun halos risks eye damage, whereas moon halos can be safely observed for extended periods. This safety factor likely contributed to moon halos receiving more cultural attention despite sun halos being more visually spectacular.

The twenty-two degree radius represents just one type of halo. Under certain conditions, ice crystals produce additional optical effects including forty-six degree halos, circumzenithal arcs, sundogs, and light pillars. Each results from light interacting with differently oriented ice crystals. The twenty-two degree halo remains the most commonly observed because it requires only random crystal orientation rather than specific alignments.

Halo brightness varies based on cirrus cloud density and ice crystal concentration. Faint halos barely visible against the night sky indicate thin cirrus veils. Brilliant halos bright enough to cast shadows suggest dense ice crystal clouds. Brighter halos generally correlate with more robust approaching weather systems, though this relationship is not absolute. Even faint halos indicate moisture at high altitude, a necessary precursor to precipitation.

The temporal relationship between halo observation and precipitation arrival depends on several factors. Wind speed at high altitudes determines how quickly cirrus clouds advance ahead of surface weather systems. Stronger upper level winds extend the warning period; slower winds compress it. The storm system's overall speed of movement affects timing. Fast moving systems reduce the interval between halo and rain; slow systems extend it.

Modern meteorological forecasting incorporates cirrus cloud observations as one data point among many. Satellite imagery reveals cirrus extent and density far beyond what ground observers can see. Weather radar tracks approaching precipitation. Computer models predict storm paths and intensities. Yet the simple act of observing a moon halo still provides useful information accessible to anyone without technological aids.

The halo omen functions anywhere on Earth where ice crystal clouds form. Arctic and Antarctic regions see frequent halos during winter months when moisture content, though low, still produces ice crystal clouds. Tropical zones observe halos less frequently but under similar meteorological circumstances. The universal physics of light refraction through hexagonal ice ensures consistent halo appearance across all latitudes and longitudes.

Photography captures moon halos more readily than many atmospheric phenomena. The halo's brightness and distinct shape make it accessible even to simple cameras. Time lapse photography reveals how halos evolve as cirrus clouds thicken and spread. Such documentation shows halos gradually blurring and fading as lower, thicker clouds move in, obscuring the moon completely before precipitation begins.

Urban light pollution complicates halo observation in cities. Artificial lighting reduces sky contrast, making faint halos difficult to detect.

However, bright halos remain visible even from urban locations, particularly when viewed from areas with less direct light exposure. The increasing prevalence of light pollution means that future generations may lose easy access to this traditional weather sign unless they deliberately seek dark sky locations.

Pilots flying at night regularly observe moon halos from above, seeing them from perspectives impossible for ground observers. From altitude, pilots may see multiple halos simultaneously if ice crystal clouds exist at different levels. These observations provide immediate information about atmospheric conditions, supplementing instrument readings with direct visual data.

The relationship between moon phase and halo visibility is straightforward: brighter moons produce more visible halos. Full or nearly full moons provide maximum illumination for ice crystal refraction. New moons produce no halos because insufficient light exists to refract. Quarter moons generate fainter halos visible only when cirrus clouds are dense enough to contain abundant ice crystals.

Lunar eclipses prevent halo formation during totality. When Earth's shadow covers the moon completely, no direct moonlight exists to refract through ice crystals. Halos may appear before or after totality if cirrus clouds are present, but the eclipse itself interrupts the phenomenon. This interaction between celestial mechanics and atmospheric optics occasionally produces memorable visual sequences as halos disappear during eclipse onset and reappear during emergence.

The persistence of moon halo folklore across cultures and centuries testifies to the phenomenon's reliability. Humans developed complex mythologies around less predictable or entirely imaginary signs, but the moon halo's consistent correlation with approaching weather embedded it in practical lore passed from generation to generation. Farmers, sailors, shepherds, travelers, all groups whose livelihoods depended on weather awareness, maintained these traditions because they worked.

Climate change affects the reliability of traditional weather signs in complex ways. Shifting jet stream patterns alter the typical progression of weather systems. Increased atmospheric moisture from warming potentially produces more cirrus clouds but also disrupts established seasonal patterns. Whether moon halos remain as reliable for precipitation forecasting in a changing climate requires ongoing observation and documentation.

The next time you notice a ring around the moon, recognize it for what it is: light from our nearest celestial neighbor bent through millions of ice crystals floating miles above your head, announcing that moisture is gathering and weather is changing. No instruments needed, no apps required. Just awareness, observation, and the ancient knowledge that the sky speaks to those who pay attention. Within forty-eight hours, in most cases, precipitation will arrive. Plan accordingly.

Chapter Three: The Green Flash

The green flash ranks among the rarest and most disputed atmospheric phenomena visible to human eyes. For centuries, sailors reported seeing a brilliant emerald spark at the exact moment the sun disappeared below the horizon. Skeptics dismissed these accounts as optical illusions, alcohol induced hallucinations, or outright fabrications. Yet the green flash is entirely real, a consequence of atmospheric refraction and light dispersion occurring under precise conditions. When observed, it signals exceptional air clarity and stable atmospheric layering, conditions that historically promised favorable sailing weather for days to come.

The phenomenon lasts only seconds at most, often just a fraction of a second. As the sun's upper edge drops below the horizon, the final visible portion sometimes flashes brilliant green instead of the expected orange or red. The color appears so vivid and unexpected that witnesses often question whether they actually saw it. This brevity and the specific conditions required for visibility mean that many people spend entire lifetimes near oceans without ever witnessing a green flash.

Understanding the green flash requires examining how Earth's atmosphere bends and separates sunlight. The atmosphere acts as a gigantic prism, refracting light passing through it at oblique angles. When the sun sits low on the horizon, its light travels through far more atmosphere than when directly overhead. This extended path through air layers of varying density bends different wavelengths by different amounts.

Light behaves as electromagnetic radiation traveling in waves. Different colors correspond to different wavelengths. Red light has the longest wavelength, around 700 nanometers. Blue and violet light have the shortest, around 400 to 450 nanometers. Green occupies the middle range at roughly 550 nanometers. When white sunlight enters the atmosphere at low angles, these wavelengths refract differently, with shorter wavelengths bending more than longer ones.

The atmosphere refracts blue and violet light most strongly, bending these short wavelengths more sharply downward than other colors. However, atmospheric molecules scatter blue and violet light intensely through a process called Rayleigh scattering. This scattering removes much of the blue and violet before it reaches an observer's eyes, the same process that makes the sky appear blue during the day. With blue and violet scattered away, green becomes the shortest wavelength reaching observers in sufficient intensity to perceive.

Red, orange, and yellow wavelengths refract less than green. At sunset, the sun's main disk appears red or orange because these longer wavelengths dominate the transmitted light after shorter wavelengths have scattered. As the sun sets, the red image of the sun disappears below the horizon first because it refracts least. The orange image follows, then yellow. The green image, refracting more strongly, remains visible a fraction of a second longer before it too drops below the horizon.

For this sequence to produce a visible green flash, atmospheric conditions must meet strict requirements. The air must be exceptionally clear, free from pollution, dust, and water vapor that would scatter green light before it reaches observers. The atmosphere must exhibit smooth, stable layering without turbulence that would distort the image. The horizon must be sharp and unobstructed, typically an ocean horizon where nothing interrupts the view between observer and the setting sun.

Temperature inversions enhance green flash visibility. When a layer of warm air sits above cooler air near the surface, light refracts more strongly through this boundary. This enhanced refraction increases the angular separation between different colored images of the sun, making the green component more distinct and easier to observe. Stable high pressure systems often create these temperature inversions, the same weather patterns that produce calm seas and clear skies.

Maritime cultures developed the belief that observing a green flash promised good weather ahead. This correlation proved accurate because the atmospheric conditions enabling green flash visibility are the same conditions associated with stable, fair weather. Clear air, temperature inversions, calm winds, these factors characterize high pressure systems that bring extended periods of pleasant weather. Sailors who saw green flashes could expect favorable conditions to persist for days.

Historical accounts of green flashes appear in ships' logs, naval records, and maritime literature dating back centuries. Captain James Cook's voyage journals mention unusual green coloration at sunset. Jules Verne's 1882 novel "Le Rayon Vert" popularized the phenomenon, though Verne added fictional romantic elements to the genuine atmospheric occurrence. These literary references helped spread knowledge of the green flash beyond seafaring communities.

Skepticism about green flash authenticity persisted until photography provided definitive proof. Early color film captured green flashes during the 1950s and 1960s, documenting the phenomenon's existence beyond subjective eyewitness testimony. Modern digital photography makes capturing green flashes relatively straightforward for observers positioned in the right locations under appropriate conditions.

The green flash actually comprises several related phenomena, all involving atmospheric refraction of sunlight. The most common type, the inferior mirage flash, appears as a green rim on the sun's upper edge for one or two seconds before disappearing. A rarer variant, the mock mirage flash, can last up to fifteen seconds under ideal atmospheric conditions. The rarest type, the green ray or green segment, appears as a vertical green beam shooting upward from the setting sun.

Elevation enhances green flash observation. Observers positioned higher above sea level see flashes more frequently and for longer durations. This occurs because greater elevation provides a more distant horizon, extending the time during which the sun appears to set. Mountain observatories near oceans offer optimal viewing locations, combining clear air at altitude with unobstructed ocean horizons.

The green flash occurs at sunrise as well as sunset, though sunset observations are more common. At sunrise, the green flash appears as the sun's first edge rises above the horizon, requiring observers to predict exactly where and when the sun will emerge. Sunset observations prove easier because the sun's position is already known, allowing observers to watch and wait for the flash as the sun descends.

Atmospheric pollution significantly reduces green flash visibility. Industrial haze, smoke, and particulate matter scatter green wavelengths before they reach observers. Urban coastal areas rarely offer conditions clear enough for green flash observation. Remote oceanic locations far from pollution sources provide the best opportunities, explaining why the phenomenon became embedded in sailing lore rather than coastal city traditions.

The curvature of Earth plays a role in green flash generation. As the sun sets, its light must travel tangent to Earth's surface, passing through maximum atmospheric thickness. This geometry maximizes refraction effects while providing the necessary viewing angle for wavelength separation to become visible. The same curvature prevents green flashes from occurring when the sun is higher in the sky.

Latitude affects green flash characteristics. Near the equator, the sun sets nearly perpendicular to the horizon, dropping quickly and producing brief flashes. At higher latitudes, the sun sets at shallower angles, taking longer to disappear and potentially producing extended green flashes. However, higher latitudes often have less stable atmospheric conditions, reducing overall flash frequency despite the favorable geometry.

Some observers report blue flashes under extraordinary conditions. These occur when atmospheric clarity is so exceptional that even blue wavelengths avoid complete scattering. Blue flashes are vanishingly rare, requiring near perfect atmospheric transparency. Their observation indicates conditions of absolutely exceptional clarity and stability.

The psychological impact of witnessing a green flash should not be underestimated. The unexpected color, the brief duration, the requirement for perfect timing, all contribute to a sense of having witnessed something magical. This emotional response reinforced the cultural significance sailors attached to the phenomenon, transforming a physical observation into an experience imbued with meaning and portent.

Deliberate green flash observation requires patience and preparation. Observers must identify locations with clear, unobstructed horizons, typically ocean coasts or high elevation sites overlooking water. Checking weather forecasts for high pressure systems and clear skies increases success probability. Arriving early allows eyes to adapt and observers to identify the exact sunset point. Watching the last few seconds of sunset with focused attention maximizes chances of catching the brief flash.

Never look directly at the sun until the very last moment before it disappears. Solar radiation can damage eyes even when filtered through atmosphere. The safe window for direct observation spans only the final seconds as the sun's last edge drops below the horizon. Prior to this, observers should watch the general sunset progression without staring directly at the solar disk.

Photography captures green flashes that happen too quickly for certain visual perception. Camera sensors respond differently than human eyes, sometimes recording green flashes invisible to observers present during the exposure. This discrepancy between visual and photographic evidence initially created confusion about the phenomenon's nature but ultimately helped confirm its physical reality.

The green flash exists entirely independent of human observation. It results from refractive physics, not perception or interpretation. The same atmospheric conditions produce identical effects whether anyone watches or not. This objective reality distinguishes it from purely subjective phenomena like pareidolia or optical illusions created by the eye itself.

Climate influences green flash frequency through effects on atmospheric clarity and stability. Arid regions with minimal moisture produce clearer air than humid zones. Stable subtropical high pressure belts offer more frequent flash opportunities than stormy temperate latitudes. Desert coasts where dry air meets ocean provide ideal combinations of clarity and horizon access.

The green flash omen retained practical value even after meteorological forecasting became sophisticated. Modern weather prediction relies on instruments, computer models, and satellite data, but direct observation of atmospheric conditions still provides immediate, local information. A visible green flash confirms that current conditions are exceptionally clear and stable, information useful for immediate planning regardless of what forecasts predict for later.

Cultural transmission of green flash knowledge varied between maritime traditions. Some sailing cultures considered witnessing the flash lucky, others treated it as matter of fact confirmation of good weather. These different cultural interpretations overlaid the same basic observation: clear air and stable conditions produced both green flashes and favorable sailing weather.

The next opportunity to witness a green flash might come tonight, tomorrow, or never, depending on location and atmospheric conditions. Those who seek this phenomenon must cultivate patience, persistence, and awareness. The reward is momentary, a brief spark of emerald light at day's end. But that moment confirms what sailors knew for centuries: the atmosphere is clear, the weather is stable, and conditions favor safe passage. In that instant, ancient knowledge and modern physics converge, both reading the same atmospheric text, both reaching the same conclusion. The sky is clear. Fair weather prevails.

Chapter Four: Red Sky at Night

Red sky at night, sailor's delight. Red sky at morning, sailor's warning. This rhyme has guided maritime decisions for centuries, passed down through generations of seafarers who staked their lives on its accuracy. The saying appears in cultures worldwide, from ancient Mediterranean civilizations to Pacific island navigators, each independently recognizing the same atmospheric correlation. Modern meteorology confirms what traditional knowledge maintained: the color of the sky at sunset and sunrise provides reliable information about approaching weather systems, particularly in temperate latitudes where prevailing winds blow from west to east.

The physics behind red skies involves light scattering through atmospheric particles. Sunlight consists of all visible wavelengths traveling together as white light. When this light enters the atmosphere, it interacts with gas molecules, water droplets, dust particles, and other suspended matter. Different wavelengths scatter differently based on particle size and light wavelength, a relationship described mathematically by Rayleigh scattering equations.

Blue light, with its short wavelength around 450 nanometers, scatters readily when encountering atmospheric molecules. This scattering sends blue light in all directions, which is why the sky appears blue during midday when the sun is overhead and light passes through minimal atmosphere. Red light, with wavelengths around 700 nanometers, scatters much less efficiently. It travels through the atmosphere more directly, penetrating farther through air before scattering significantly.

At sunset and sunrise, sunlight must travel through far more atmosphere to reach observers than at midday. The low angle means light passes through perhaps thirty times more air than when the sun sits directly overhead. This extended path through the atmosphere scatters away most blue and green wavelengths before they reach observers' eyes, leaving primarily red, orange, and yellow light visible.

The intensity and extent of red coloration depend on atmospheric particle content. Clean, dry air produces modest reddening. Air containing more particles, whether dust, smoke, water vapor, or pollution, scatters light more intensely and removes shorter wavelengths more completely, producing deeper red hues. However, too much particulate matter, such as thick clouds or heavy pollution, can block light entirely rather than just scattering it selectively.

The "red sky at night" portion of the saying relates to weather system movement in temperate zones. Prevailing winds in these latitudes generally flow from west to east, driven by the rotation of Earth and atmospheric circulation patterns. Weather systems, including high and low pressure areas, move with these winds from west to east. When observers see a red sunset to their west, they are looking through clear air toward the setting sun, indicating that high pressure and fair weather dominate the western sky.

High pressure systems bring descending air that inhibits cloud formation. This descending air compresses and warms, reducing relative humidity and creating clear conditions. The clear western sky at sunset appears red because sunlight passes through dry air relatively free of water droplets that would otherwise scatter all wavelengths and create white or gray appearance. The red sky signals that this clear, high pressure weather currently to the west will arrive at the observer's location as prevailing winds push it eastward overnight.

The correlation proves remarkably reliable in temperate maritime zones. Studies of the saying's accuracy in regions like the North Atlantic and Mediterranean show success rates between sixty and seventy percent. The saying works best during fall, winter, and spring when large scale weather systems dominate atmospheric patterns. Summer brings more localized convective weather that develops independently of prevailing flow, reducing the saying's reliability.

Red sky in the morning presents the opposite scenario. Dawn breaks in the east, so observers seeing red sunrise are looking eastward through clear air. The clear conditions lie to the east, meaning they have already passed the observer's location during the night, moving eastward with prevailing winds. To the west, where weather will come from during the day, conditions remain unknown but are likely deteriorating. The clear air that produced the red sunrise will continue moving east, while cloudy or stormy weather from the west approaches.

Additionally, morning red skies often indicate high altitude cirrus clouds to the west catching first light from the rising sun. These ice crystal clouds appear reddish when illuminated from below by the rising sun shining through the lower atmosphere. Cirrus clouds typically precede warm fronts and approaching low pressure systems by twelve to forty-eight hours, providing advance warning of deteriorating conditions. The red morning sky thus combines two signals: clear air to the east where weather has already been, and high clouds to the west where weather is approaching.

Biblical references demonstrate the saying's ancient pedigree. Matthew 16:2-3 records Jesus citing this weather lore: "When it is evening, you say, it will be fair weather, for the sky is red. And in the morning, it will be stormy today, for the sky is red and threatening." This passage confirms that Mediterranean cultures in the first century recognized and relied on sky color for weather prediction.

Greek and Roman writers documented the same observations. Theophrastus, a student of Aristotle writing around 300 BCE, included red sky interpretations in his meteorological works. Virgil's Georgics, a didactic poem about farming written in the first century BCE, contains detailed weather lore including references to sunset and sunrise coloration as predictive tools.

The global distribution of similar sayings suggests independent discovery across multiple maritime cultures. Pacific island navigators developed identical interpretations of red skies despite having no contact with European or Mediterranean traditions. Chinese weather lore includes parallel sayings linking red sunsets to fair weather. This convergence across isolated cultures indicates that the correlation reflects genuine atmospheric patterns observable anywhere that prevailing winds exist.

Latitude affects the saying's applicability. Tropical regions lack strong prevailing westerly winds, instead experiencing trade winds and seasonal monsoon patterns. Weather in the tropics develops more through local convection than through traveling weather systems. The red sky saying consequently works less reliably near the equator. Conversely, temperate zones between roughly thirty and sixty degrees latitude see the highest correlation between sky color and subsequent weather.

Exceptional red sunsets, those producing vivid crimson or orange skies, often follow volcanic eruptions that inject fine particles into the stratosphere. The 1883 eruption of Krakatoa produced spectacular red sunsets visible globally for years as volcanic aerosols circulated through the upper atmosphere. These particles, sized perfectly for scattering blue light while transmitting red wavelengths, created sunsets so intense that fire departments responded to false alarms from citizens who thought the red sky indicated distant fires.

Air pollution creates similar effects on a smaller scale. Urban and industrial areas often display enhanced red sunsets due to particulate matter in the lower atmosphere. However, this pollution also complicates weather prediction based on sky color because the reddening may result from local pollution rather than atmospheric moisture content or pressure patterns. The cleaner air over oceans makes the saying more reliable for maritime applications than for landlocked industrial regions.

Seasonal variations influence sky color intensity. Winter air in temperate zones tends to be drier, producing clearer conditions when high pressure dominates. The lower sun angle during winter also means sunlight travels through more atmosphere even at midday, enhancing color effects. Summer brings higher humidity and more variable conditions, sometimes producing red skies that don't follow the traditional weather correlation.

Photography and modern color measurement confirm that human perception of red skies accurately reflects physical reality. Spectrophotometers measuring wavelength distribution in sunset and sunrise light show predominance of longer wavelengths exactly as observers report. The phenomenon is not subjective interpretation but measurable physical fact resulting from known atmospheric processes.

The emotional and aesthetic response to red skies likely reinforced their cultural significance. Vivid sunsets inspire awe and attention, making them memorable and ensuring observations get discussed and transmitted. Less dramatic sky colors might go unnoticed, but spectacular red sunsets demanded attention and comment, facilitating the accumulation of observational data across generations.

Mariners developed nuanced interpretations beyond the basic rhyme. Deep blood red skies suggested different conditions than pale pink ones. Purple tinged sunsets indicated specific moisture levels. Orange rather than red coloration meant different particle sizes in the atmosphere. These refinements required years of observation to master but provided additional predictive information for experienced sailors.

The reliability of red sky predictions decreased with distance from oceans. Interior continental locations experience different weather patterns than coastal areas. Mountains disrupt prevailing winds. Large lakes create localized weather effects. The saying evolved in maritime contexts and works best there, though it retains some utility inland where large scale weather systems still generally move west to east.

Modern meteorology incorporates sky color observation as supporting data rather than primary forecasting tool. Satellite imagery, radar, computer models, and instrument measurements provide far more detailed information than visual observation alone. However, sky color remains accessible to anyone anywhere, requiring no technology or expertise beyond simple attention to the world overhead.

Climate change may affect the traditional accuracy of weather lore like the red sky saying. Shifting jet stream patterns alter typical weather system tracks. Changing precipitation patterns affect atmospheric moisture content. Increased wildfire frequency injects more smoke particles into the air. Whether these changes significantly degrade the correlation between sky color and subsequent weather requires ongoing observation and documentation.

Indigenous weather prediction systems incorporated red sky observations within broader frameworks of environmental monitoring. Sky color combined with wind direction, cloud types, animal behavior, and seasonal patterns to create comprehensive forecasting methods. No single sign determined predictions; rather, multiple indicators together built reliable forecasts. This holistic approach recognized that atmospheric conditions manifest through numerous observable phenomena simultaneously.

The persistence of the red sky saying in popular culture, appearing in movies, songs, and literature, ensures its transmission even to populations with no maritime experience. This cultural survival means that basic atmospheric literacy persists in societies increasingly disconnected from direct weather dependence. The saying serves as gateway to deeper understanding of how observation and pattern recognition can yield practical knowledge.

The next time you witness a vivid red sunset, understand what you are seeing: light from our star filtered through miles of atmosphere, its shorter wavelengths scattered away by air molecules and suspended particles, leaving red light to paint the sky. That color tells a story about atmospheric conditions, about air clarity and moisture content, about pressure systems and wind patterns. In temperate latitudes, more often than not, that red evening sky promises pleasant weather tomorrow. The sailors knew. The farmers knew. The knowledge remains available to anyone who looks up and pays attention. The sky still speaks the same language it always has. We need only remember how to listen.

Chapter Five: The Fallstreak Hole

A circular gap suddenly appears in an otherwise uniform cloud layer, as if something punched through from above. Radiating streamers of ice crystals fall from the hole's edges, creating the appearance of a vortex or portal. Witnesses photograph these formations frantically, posting images online with speculation about UFOs, government weather manipulation, or supernatural intervention. Yet the fallstreak hole, also called a hole punch cloud or cavum, results from entirely natural atmospheric physics involving supercooled water and ice nucleation. Historically, these formations were interpreted as omens of approaching ice storms, a connection that modern meteorology confirms has basis in atmospheric conditions.

Fallstreak holes form in mid level altocumulus or cirrocumulus clouds, typically between 8,000 and 20,000 feet altitude. These clouds consist of supercooled water droplets, liquid water existing below freezing temperature without crystallizing into ice. Water can remain liquid down to approximately negative forty degrees Celsius if no ice nuclei are present to trigger crystallization. The droplets exist in metastable equilibrium, liquid but poised to freeze instantly if disturbed.

The instability of supercooled water creates the conditions necessary for fallstreak hole formation. When a portion of the cloud layer begins freezing, the ice crystals that form grow rapidly at the expense of surrounding water droplets. Ice crystals and liquid water droplets cannot coexist in equilibrium at the same temperature. Water molecules preferentially move from liquid phase to ice phase through vapor transfer, causing ice crystals to grow while water droplets evaporate.

This process, called the Bergeron process after Swedish meteorologist Tor Bergeron who described it in the 1930s, drives much precipitation formation in clouds. Ice crystals have lower vapor pressure than water droplets at the same temperature, meaning water molecules more readily attach to ice surfaces than remain in liquid droplets. As ice crystals grow, they deplete water vapor in their immediate vicinity, causing nearby droplets to evaporate to replenish the vapor. The net effect transfers water from droplets to ice crystals.

In a fallstreak hole, this process initiates locally in a small area of the cloud layer and propagates outward. The initial ice formation might be triggered by an aircraft climbing or descending through the cloud layer. As the aircraft passes through supercooled clouds, aerodynamic cooling around the wings and fuselage, particularly at wing tips where pressure drops sharply, can lower air temperature enough to trigger ice nucleation. The propeller wash or jet exhaust also provides ice nuclei through particle emissions.

Once ice crystals form, they begin growing and falling. As they fall, they pass through more supercooled cloud below, triggering additional freezing. The process expands radially from the initiation point, creating the characteristic circular or elliptical hole. The ice crystals falling from the hole's edges create virga, visible streaks of precipitation that evaporate before reaching the ground. These trailing streamers give fallstreak holes their distinctive appearance, resembling jellyfish or spaceships depending on observer perspective.

The holes expand to remarkable sizes, sometimes exceeding thirty miles in diameter. Satellite imagery captures their extent far better than ground observations. From above, fallstreak holes appear as nearly perfect circles punched through otherwise uniform cloud layers, dramatic disruptions of regular patterns. Multiple holes can form in the same cloud system if conditions favor ice nucleation at several locations.

The atmospheric conditions required for fallstreak hole formation correlate with approaching winter weather systems. Mid level clouds containing abundant supercooled water typically exist ahead of warm fronts or within broad areas of lift associated with upper level troughs. These same meteorological patterns often precede ice storms, freezing rain, and heavy snow. Populations observing fallstreak holes could reasonably interpret them as warnings of icy precipitation within twenty four to forty eight hours.

Historical accounts describing holes in clouds exist throughout recorded history, though interpretations varied dramatically from modern understanding. Medieval European chronicles occasionally mention circular gaps appearing in overcast skies, sometimes interpreted as divine signs or omens of plague. Norse sagas reference similar phenomena in their descriptions of portentous weather. These accounts likely describe fallstreak holes, though cultural interpretation overlaid the observations with supernatural significance.

The UFO explanation for fallstreak holes gained traction in the mid twentieth century as aircraft traffic increased and created more frequent hole formations. Witnesses unfamiliar with cloud physics saw circular gaps with trailing streamers and reasonably concluded that some craft had punched through the clouds. The geometric regularity and dramatic appearance matched popular conceptions of alien technology. News media periodically reported these sightings, often without consulting meteorologists who could have provided natural explanations.

Scientific understanding of fallstreak holes developed gradually through the twentieth century. Early aviation encounters with supercooled clouds revealed their instability. Pilots flying through these clouds sometimes observed ice forming on aircraft surfaces and clouds dissipating in their wake. Meteorologists connected these observations to the Bergeron process and recognized that aircraft could trigger large scale cloud modifications.

Controlled studies in the 1980s and 1990s confirmed the aircraft trigger mechanism. Researchers tracked aircraft flight paths through satellite imagery of developing fallstreak holes, finding strong correlations between aircraft passages and hole initiation. The timing matched expectations: holes appeared minutes after aircraft penetrated cloud layers, with the delay corresponding to the time required for ice nucleation and crystal growth to become visible.

Not all aircraft passages through supercooled clouds produce fallstreak holes. The clouds must be in a specific state, with temperatures in the critical range where supercooled water is abundant but unstable. Too warm, and water droplets remain stable. Too cold, and ice crystals already dominate. The narrow temperature window, typically between negative ten and negative twenty degrees Celsius, determines where fallstreak holes can form.

The type of aircraft also matters. Propeller driven aircraft produce more effective ice nucleation than jets because propeller wash creates more turbulence and mixes more exhaust particles into clouds. However, jets flying at specific angles, particularly during climbs or descents, can also trigger holes through aerodynamic effects alone without relying on exhaust particles.

Military aviation occasionally produced spectacular fallstreak formations during training exercises. Multiple aircraft flying in formation through supercooled cloud layers simultaneously triggered ice nucleation at many points, creating complex patterns of overlapping holes. Witnesses on the ground seeing these formations naturally assumed technological rather than meteorological causes, not understanding the atmospheric physics involved.

The time of day affects fallstreak hole visibility. Holes forming at sunrise or sunset may be illuminated dramatically when the sun shines through the gap and catches falling ice crystals at angles that don't illuminate the surrounding cloud. This lighting creates spectacular visual effects, with the hole appearing to glow while surrounding clouds remain shadowed. Such displays amplify the mysterious appearance and fuel unconventional explanations.

Geographic patterns in fallstreak hole reports reflect both flight paths and atmospheric conditions. Regions under major air traffic corridors see more holes than remote areas with minimal aviation. However, the clouds must also support supercooled water, meaning temperature and moisture conditions must align. Continental interiors in winter provide ideal conditions, combining necessary atmospheric states with heavy air traffic.

Climate affects fallstreak hole frequency through impacts on cloud characteristics and atmospheric stability. Warming temperatures may shift the altitude bands where supercooled clouds form. Changes in atmospheric circulation patterns alter where and when conditions favor these cloud types. The relationship between climate trends and fallstreak hole frequency remains an active research area.

Satellite technology revolutionized fallstreak hole documentation. Before comprehensive satellite coverage, holes were observed primarily from the ground with no sense of their full extent or frequency. Satellite imagery reveals that fallstreak holes occur far more commonly than ground observations suggested, with dozens forming daily across regions with appropriate conditions. This discovery highlighted how limited ground based observation can be for atmospheric phenomena.

The longevity of fallstreak holes varies. Some persist for hours, slowly expanding until the entire cloud layer dissipates or transforms to ice crystals. Others close within minutes as surrounding cloud moves into the gap, though the ice formation process often continues, just becoming less visually distinct. Wind shear can distort circular holes into elongated ellipses or irregular shapes as different altitudes experience different wind speeds.

The ice crystals falling from fallstreak holes sometimes reach the ground if the air below is sufficiently cold and humid. More commonly, they sublimate or evaporate in warmer air at lower altitudes, creating virga visible as streaks beneath the hole. This precipitation that never reaches the surface still plays a role in atmospheric moisture redistribution, moving water from cloud level downward through evaporation and vapor transport.

Educational efforts about fallstreak holes aim to reduce misidentification as unexplained phenomena. Meteorological agencies occasionally issue public information when holes appear over populated areas, explaining the natural processes involved. These efforts meet mixed success, as dramatic visual phenomena tend to resist prosaic explanations in popular imagination. The desire for mystery often overwhelms scientific clarity.

Photography of fallstreak holes benefits from wide angle lenses that capture their full extent. Time lapse photography reveals their formation and evolution, showing how holes grow from small disturbances into large scale cloud modifications. These visual records provide valuable data for refining understanding of ice nucleation processes and cloud physics.

The omen interpretation of fallstreak holes as ice storm predictors contains legitimate meteorological insight. The clouds that produce holes exist in atmospheric conditions that often precede freezing precipitation. Supercooled water at mid levels indicates abundant moisture and temperatures hovering near freezing through a deep atmospheric layer. These conditions frequently transition to ice storms, freezing rain, or heavy snow as weather systems evolve. Observers seeing fallstreak holes and preparing for icy conditions were responding appropriately to visible atmospheric signals.

Modern weather forecasting identifies these same conditions through instrument measurements and computer modeling. Forecasters monitor temperature profiles, moisture content, and cloud characteristics to predict freezing precipitation. A visible fallstreak hole provides immediate confirmation that atmospheric conditions match those associated with ice storm potential. The hole doesn't cause the ice storm but indicates that the atmosphere exists in a state conducive to icy precipitation.

The next time you see a circular gap punched through clouds with trailing streamers falling from its edges, understand the physics at work. Supercooled water, that strange state where liquid persists below freezing, has been disturbed, triggered into crystallization by passing aircraft or natural turbulence. Ice crystals grow and fall, clearing a hole through the cloud layer. The same conditions that allow this process also favor freezing precipitation. The ancients saw omens. We see atmospheric physics. Both perspectives recognize the same truth: the sky announces coming ice through visible signs. Whether interpreted as warning or wonder, the message remains. Cold weather approaches. Ice will form. Prepare accordingly.

Chapter Six: Comet Fever

Throughout history, comets blazing across the night sky triggered mass hysteria, religious fervor, and waves of inexplicable illness. Populations interpreted these celestial visitors as harbingers of disaster, divine judgment, or poisonous miasmas threatening to corrupt Earth's atmosphere. From ancient Rome to medieval Europe to nineteenth century America, comet appearances coincided with documented outbreaks of strange maladies, bizarre behavior, and collective panic. Modern scholarship reveals a disturbing correlation: many historical comet panic episodes align precisely with periods of widespread lead poisoning from contaminated water supplies, food preservation methods, and industrial practices. The omens were real, but their source lay not in the heavens but in the toxic metals populations unknowingly consumed daily.

Comets appear as fuzzy points of light in the night sky, sometimes developing spectacular tails stretching across substantial portions of the celestial sphere. These icy bodies originate from the outer solar system, entering the inner solar system on elliptical orbits that bring them near the sun. Solar radiation heats the comet's icy surface, causing sublimation that releases gas and dust forming the characteristic tail. The tail always points away from the sun, pushed by solar wind and radiation pressure.

Ancient astronomers distinguished comets from planets and stars through their movement and appearance. Planets moved along predictable paths. Stars maintained fixed relative positions. Comets appeared unpredictably, moved through constellations, and then vanished, sometimes never returning within human lifetimes. This unpredictability combined with their dramatic appearance made comets objects of fear and fascination across all cultures.

Chinese astronomical records dating back over two thousand years meticulously documented comet appearances. These records describe the comet's position, movement, color, tail length, and any significant events occurring during the apparition. Many entries note that comet appearances coincided with deaths of emperors, military defeats, famines, or epidemics. Similar patterns appear in European chronicles, Middle Eastern texts, and records from other literate civilizations.

The comet of 44 BCE appeared shortly after Julius Caesar's assassination. Roman citizens interpreted the comet as Caesar's soul ascending to join the gods. The dictator's heir, Octavian, exploited this interpretation for political benefit, commissioning art depicting Caesar's transformation into a deity. The comet's appearance during this politically turbulent period became permanently associated with Caesar's death in historical memory.

Medieval Europe viewed comets with pure terror. The Catholic Church taught that comets signaled God's anger and impending punishment. When comets appeared, populations flocked to churches for confession and prayer. Authorities proclaimed days of fasting and penance. Some communities expelled sinners or conducted mass executions of criminals in desperate attempts to appease divine wrath. The comet of 1066 appeared shortly before the Norman conquest of England, forever linking the celestial event with that watershed moment in English history.

The Great Comet of 1680 produced widespread panic across Europe and colonial America. This exceptionally bright comet remained visible for months, its tail stretching across much of the sky. Contemporary accounts describe populations gripped by terror, convinced that the comet's poisonous vapors would corrupt the air and cause plague. Some people sealed their windows and doors, others fled cities for rural areas. Medical practitioners reported surges in patients complaining of headaches, confusion, weakness, and digestive problems.

These symptoms match lead poisoning precisely. Lead toxicity causes neurological effects including headache, confusion, irritability, and cognitive impairment. Gastrointestinal symptoms include abdominal pain, constipation, and nausea. More severe poisoning produces seizures, paralysis, and death. Historical populations exposed to chronic low level lead poisoning would experience these symptoms continually, but might only recognize and report them during periods of heightened anxiety such as comet appearances.

Lead contamination pervaded daily life in historical societies. Romans used lead extensively for water pipes, cooking vessels, and food preservation. Lead acetate, called "sugar of lead" for its sweet taste, was deliberately added to wine as a preservative and flavor enhancer. Roman aristocrats consumed enormous quantities of lead through their diet and drinking water. Skeletal remains from the Roman period show lead levels far exceeding modern safety thresholds.

Medieval and early modern Europe continued these practices. Pewter vessels, which contain lead, were standard for eating and drinking. Lead glazes on pottery leached into acidic foods and beverages. Lead compounds were used in cosmetics, medicines, and food adulterants. Urban water supplies flowing through lead pipes accumulated dangerous concentrations. House paint contained lead, creating dust that residents inhaled and ingested.

The correlation between comet appearances and reported illness likely reflects heightened attention to existing symptoms rather than new disease outbreaks. Populations chronically poisoned by environmental lead experienced continuous low grade symptoms. During ordinary times, these symptoms might be dismissed or attributed to other causes. When a comet appeared and authorities proclaimed atmospheric contamination, people noticed their symptoms and attributed them to celestial influence.

This psychological mechanism, called attention bias, causes people to notice phenomena that align with expectations while ignoring contradictory information. If you believe a comet's vapors cause illness, you will notice any symptoms you experience during the comet's visibility and attribute them to celestial influence. Symptoms present before the comet appeared or persisting afterward receive less attention because they don't fit the narrative.

The 1835 appearance of Halley's Comet triggered widespread panic in the United States. Newspapers published sensational articles claiming that the comet's tail would engulf Earth, poisoning the atmosphere with deadly gases. Entrepreneurs sold "comet pills" and other protective remedies. Some accounts describe people fainting from fear or barricading themselves indoors. Medical records from this period document increased reports of anxiety, insomnia, headaches, and digestive complaints.

American society in 1835 had substantial lead exposure. Lead was used in paint, glazed pottery, pewter utensils, water pipes, and various industrial processes. Urban populations, particularly in growing industrial cities, faced higher exposure than rural residents. The very symptoms attributed to the comet matched those of lead toxicity, but no one connected environmental metal exposure to their health complaints.

Halley's Comet returned in 1910, again producing panic despite a century of astronomical advances. Scientists had determined that comets posed no threat to Earth, that their tails consisted of diffuse gas too thin to affect our atmosphere. Nevertheless, newspapers published alarming articles, and some populations panicked. Reports emerged of people sealing homes against poisonous gases, others committing suicide to avoid a prolonged, agonizing death from comet induced asphyxiation.

The 1910 panic occurred during an era of massive industrial lead use. Leaded gasoline had not yet been introduced, but lead paint, lead water pipes, lead soldered cans, and lead glazed pottery remained ubiquitous. Factory workers faced intense occupational exposure. Children chewed on lead painted toys and furniture. Urban populations breathed air contaminated with lead from various industrial sources. Chronic lead poisoning affected millions, though medical understanding remained limited.

The psychological phenomenon of mass psychogenic illness helps explain how comet fears could trigger waves of reported symptoms. When a population believes it has been exposed to a toxin or pathogen, susceptible individuals may develop genuine physical symptoms triggered by anxiety and social contagion. These symptoms are real, not imaginary, but they arise from psychological rather than toxicological causes. In populations already experiencing actual lead induced symptoms, the addition of psychogenic effects would amplify overall morbidity dramatically.

Indigenous populations in the Americas also observed comets and incorporated them into cultural narratives, though interpretations varied from European traditions. Some groups viewed comets as spiritual messengers or symbols of transformation. Others saw them as omens of warfare or political change. These interpretations generally lacked the toxic contamination narrative prevalent in European thought, possibly because indigenous technologies involved less lead exposure.

The scientific understanding of comets advanced substantially through the seventeenth and eighteenth centuries. Edmond Halley's prediction that the comet observed in 1682 would return in 1758 proved correct, demonstrating that comets followed natural laws like other celestial bodies. This breakthrough shifted comet perception among educated classes from supernatural omens to predictable astronomical phenomena. However, popular fear persisted, particularly in populations with limited scientific education.

Modern spectroscopy allows detailed analysis of comet composition. Comets consist primarily of water ice, frozen carbon dioxide, methane, ammonia, and various organic compounds. When heated by solar radiation, these materials sublimate and release gases that form the visible coma and tail. While some comet gases could theoretically be toxic at high concentrations, the extreme diffusion of comet material means Earth's passage through a comet tail has no detectable atmospheric effects.

The 1996 appearance of Comet Hyakutake and 1997 appearance of Comet Hale Bopp produced no mass panic, though some fringe groups promoted apocalyptic interpretations. The Heaven's Gate cult famously committed mass suicide based on beliefs involving Hale Bopp. This tragedy demonstrated that comet related fears persist in some populations despite widespread scientific literacy. However, the absence of broader panic shows that cultural attitudes have shifted substantially from historical patterns.

Lead exposure has declined dramatically in developed nations over the past fifty years. The elimination of leaded gasoline, lead paint, and lead pipes, combined with environmental regulations and awareness, reduced population level lead exposure substantially. This decline correlates with improved cognitive outcomes, reduced crime rates, and better overall health. The removal of chronic low level lead poisoning eliminated one factor that may have amplified comet panic in historical populations.

However, lead exposure remains a serious problem in many parts of the world. Developing nations often lack regulations controlling lead use. Informal battery recycling operations release massive lead contamination. Some traditional medicines and cosmetics contain lead. Lead paint persists in older buildings. These populations face the same chronic poisoning that affected historical European and American societies, though without the comets to blame for their symptoms.

The historical connection between comets and illness reveals how humans search for explanations when faced with mysterious phenomena. Chronic lead poisoning produced neurological and physical symptoms that population had no framework for understanding. When dramatic celestial events occurred simultaneously, the temporal correlation suggested causation. This pattern matching helped communities make sense of their suffering, even though the attributed cause was entirely wrong.

Modern understanding recognizes both the astronomical reality of comets and the toxicological reality of lead poisoning as separate phenomena. Comets follow gravitational laws, traveling through space on predictable orbits. Lead poisoning follows dose response relationships, damaging health through well understood chemical mechanisms. The historical conflation of these unrelated phenomena resulted from ignorance, not from any actual connection.

The comet omen, then, was both false and true. False because comets themselves posed no threat. True because comet appearances occurred during periods when populations suffered real health problems from environmental contamination. The celestial visitors served as screens onto which societies projected their fears about very real but poorly understood threats to their wellbeing. The sky provided the spectacle. The earth, specifically the lead contaminating soil, water, and food, provided the actual danger.

The next time a bright comet graces our skies, we can appreciate its beauty without fear. Modern science has revealed comets as ancient wanderers from the solar system's edge, frozen relics from planetary formation, messengers carrying information about conditions billions of years ago. They threaten nothing. But the historical lesson remains valuable: environmental toxins can produce population level health effects that societies struggle to explain. Today we worry about microplastics, PFAS chemicals, air pollution. Future generations may look back at our era and recognize correlations we currently miss, seeing connections between environmental exposures and health outcomes that seem mysterious to us now. The sky showed no omens of lead poisoning. The true warning signs existed in pipes, paint, and pottery all along.

Chapter Seven: The Blood Moon

The moon ascending blood red over the horizon transfixes observers with primal unease. This crimson orb, so unlike the familiar silver white sphere of ordinary nights, has triggered prophetic movements, apocalyptic fears, and military decisions throughout recorded history. Ancient chroniclers described blood moons preceding battles, accompanying plagues, and marking the deaths of kings. Modern atmospheric science reveals that the blood moon results from particulate matter suspended in Earth's atmosphere, the same aerosols produced by massive fires, volcanic eruptions, and dust storms. The omen carried truth: blood moons signal fire, whether from burning forests, erupting volcanoes, or the conflagrations of war itself.

The blood moon effect occurs when the moon appears near the horizon and atmospheric conditions scatter shorter wavelengths of light while allowing longer red wavelengths to reach observers' eyes. This is fundamentally the same process that reddens sunsets, but applied to moonlight reflected from the lunar surface. The moon itself undergoes no change; the coloration arises entirely from light passing through Earth's atmosphere before reaching viewers.

Under normal atmospheric conditions, the moon near the horizon appears slightly orange or yellow due to the extended path length through air. Light from the rising or setting moon must travel through perhaps thirty times more atmosphere than when the moon sits directly overhead. This extended path scatters away blue and green wavelengths through Rayleigh scattering, leaving predominantly red and orange light to reach observers. The effect produces the familiar warm tones of moonrise and moonset.

When the atmosphere contains elevated concentrations of particulate matter, this reddening intensifies dramatically. Smoke particles, volcanic ash, dust, and other aerosols scatter light more efficiently than clean air. These particles particularly affect shorter wavelengths, removing blue and green light almost completely while allowing red wavelengths to pass. The result is a moon that appears deep red, orange red, or even brownish red, far more intensely colored than normal horizon moonrise.

The size of particles affects which wavelengths scatter most effectively. Particles roughly the size of light wavelengths scatter that color most efficiently. Smoke particles from wildfires, typically between 0.4 and 1.0 micrometers in diameter, scatter blue and green light strongly while having less effect on red light around 0.7 micrometers wavelength. Volcanic ash particles vary more widely in size but often include substantial populations in the optimal range for visible light scattering.

Major volcanic eruptions inject enormous quantities of sulfur dioxide into the stratosphere, where it converts to sulfuric acid droplets. These aerosols can persist for years, circling the globe at altitudes above weather patterns. The 1883 eruption of Krakatoa produced spectacular atmospheric effects including blood red moons visible worldwide for more than two years afterward. Historical chronicles following major volcanic events consistently describe unusual lunar and solar coloration.

The year 536 CE represents one of the most catastrophic periods in recorded history. Tree ring data, ice cores, and historical accounts document that a massive volcanic eruption, likely in Iceland or the tropics, ejected so much material into the atmosphere that global temperatures dropped significantly. Byzantine historian Procopius described the sun giving forth its light without brightness during this year, like the moon during its entire course. Chinese records note that the sun shone dimly, like a lamp, and that crops failed across the empire.

The atmospheric loading from this eruption would have produced intensely red moons whenever the moon rose near the horizon. Mediterranean and Asian chroniclers describe the moon taking on colors of blood and copper. These observations coincided with crop failures, famine, plague, and social collapse across multiple civilizations. The blood moon accurately signaled fire, in this case volcanic fire injecting massive quantities of particulate matter into the global atmosphere.

Wildfire smoke produces similar effects on smaller geographic scales. Major forest fires burning millions of acres generate smoke plumes that travel thousands of miles downwind. These plumes contain fine particulate matter that scatters light efficiently. Populations downwind from massive fires observe blood red moons rising through smoke laden skies. The correlation between blood moons and fire is direct and observable: the moon appears red specifically because fire has contaminated the atmosphere.

The Great Fire of Rome in 64 CE burned for six days, destroying much of the city. Historical accounts describe the smoke as so thick that day turned to night. When the moon rose through this smoke, it would have appeared intensely red. Nero famously blamed Christians for the fire, initiating persecutions that became foundational to Christian martyrdom narratives. Whether Nero observed blood red moons during this period remains undocumented, but the atmospheric conditions certainly existed.

Medieval siege warfare involved burning cities and fields. Defending forces and attacking armies alike used fire as a weapon. Chronicles of major sieges often mention unusual sky phenomena, including red moons and suns. The atmospheric particulate matter from burning settlements, forests, and crops would have produced exactly these effects. Blood moons during wartime served as literal visual confirmation that fire was consuming the landscape.

Native American traditions across North America associate red moons with fire and danger. Tribes inhabiting fire prone ecosystems would have regularly observed blood moons during wildfire season. The correlation between red lunar coloration and active fires in the region would have been obvious to populations living in close relationship with their environment. The lunar omen accurately predicted fire danger because the red coloration resulted from existing fires.

Australian Aboriginal cultures developed similar associations. Australia's fire adapted ecosystems burn regularly, and Aboriginal peoples used fire extensively for land management. Blood moons appearing during fire season would have been common observations. Dreamtime stories incorporating red celestial phenomena likely encode information about fire behavior, atmospheric conditions, and seasonal patterns.

Lunar eclipses can also produce blood red moons through a different mechanism. During a total lunar eclipse, Earth blocks direct sunlight from reaching the moon. However, sunlight passing through Earth's atmosphere refracts into Earth's shadow, illuminating the moon with a reddish glow. The same atmospheric scattering that produces red sunsets colors this refracted light. The more particulate matter in the atmosphere, the redder the eclipsed moon appears.

The distinction between atmospheric blood moons and eclipse blood moons matters for omen interpretation. Atmospheric blood moons, caused by smoke or ash, signal active fires or recent volcanic eruptions. Eclipse blood moons follow predictable astronomical schedules unrelated to fire. However, pre-scientific observers might not distinguish clearly between these two causes. Both produced red moons, both inspired fear, and both became incorporated into prophetic traditions.

Biblical references to blood moons appear in both Old and New Testaments. The Book of Joel describes the sun turning to darkness and the moon to blood before the coming of the great and terrible day of the Lord. The Book of Revelation references the moon becoming like blood during apocalyptic events. These passages likely draw on observations of actual blood moons, whether from atmospheric particulates or eclipses, and incorporate them into eschatological frameworks.

The tetrad of lunar eclipses in 2014 and 2015 triggered modern apocalyptic speculation. Some Christian groups promoted the idea that these four blood moons signaled prophetic fulfillment and impending end times. The eclipses occurred on schedule as astronomers predicted, appeared red as expected due to atmospheric refraction, and the world continued unchanged afterward. This episode demonstrates how eclipse blood moons can still inspire prophetic interpretation despite complete astronomical understanding.

Atmospheric blood moons carry genuine information about environmental conditions. A moon appearing intensely red signals elevated particulate concentrations, which almost always indicates either fire smoke or volcanic emissions. In pre-industrial societies, fire represented serious danger. Uncontrolled wildfires threatened settlements, crops, and forests. Cities burned readily, constructed largely of flammable materials. A blood moon warning of fire was a warning worth heeding.

The severity of reddening correlates roughly with particulate concentration. Pale orange indicates modest particulate loading. Deep red suggests heavy contamination. Brownish red indicates extremely high aerosol concentrations. Experienced observers could potentially gauge fire severity or volcanic emission intensity from lunar coloration, though no historical sources document systematic calibration of this relationship.

The duration of blood moon phenomena provides additional information. A single night of red moonrise suggests a localized fire or transient smoke plume. Weeks or months of consistently red moons indicate sustained atmospheric loading from major volcanic eruptions or regional fire complexes. The 1815 eruption of Mount Tambora produced the "Year Without a Summer" in 1816, during which moons appeared red for months as volcanic aerosols circled the globe.

Modern air quality monitoring uses particulate matter measurements to assess health risks and visibility impairment. Fine particulate matter smaller than 2.5 micrometers, designated PM2.5, poses the greatest health hazards as these particles penetrate deep into lungs. These are also the particles most effective at scattering visible light and producing blood moon effects. A blood moon visible tonight correlates with elevated PM2.5 concentrations and corresponding air quality warnings.

Climate change increases wildfire frequency and intensity in many regions. Longer fire seasons, more extreme heat, and altered precipitation patterns create conditions favorable for large scale burning. Blood moons may become more common in areas previously unaffected by major wildfire smoke. This shift transforms a relatively rare omen into a regular occurrence, changing its cultural significance.

Volcanic eruptions remain unpredictable on human timescales. Centuries can pass between major eruptions producing global atmospheric effects. However, the Earth hosts numerous active volcanoes, and eruptions severe enough to create regional blood moon phenomena occur somewhere almost annually. Global interconnection means that volcanic events anywhere can now be immediately recognized for what they are rather than misattributed to supernatural causes.

The photograph documentation of blood moons proves challenging because cameras process color differently than human eyes. Camera sensors may not reproduce the intense red that observers perceive visually. Exposure settings that capture the moon's brightness often fail to record the atmospheric color accurately. This technical limitation means that written descriptions sometimes better convey blood moon appearance than photographs.

Urban light pollution interferes with blood moon observation. Artificial lighting reduces sky contrast, making subtle coloration less apparent.

However, intense blood moons remain visible even from cities. The most dramatic blood moons overcome light pollution through sheer color intensity, appearing as distinctly red orbs impossible to miss.

The emotional impact of blood moons transcends rational explanation. Even viewers understanding the physical causes may feel visceral unease when confronting a red moon rising over darkened landscape. This response likely reflects deep evolutionary programming associating fire with danger. Humans evolved in savanna environments where fire represented existential threat. The visual signal of a blood red celestial body may trigger instincts developed over millions of years.

Cultural narratives around blood moons vary but share common themes of danger, transformation, and divine communication. Western traditions emphasize apocalyptic interpretations. Eastern traditions often incorporate blood moons into dynastic omen systems predicting political change. Indigenous traditions typically connect them more directly to fire and environmental conditions. These varied interpretations all respond to the same physical phenomenon, filtered through different cultural frameworks.

The predictive value of blood moons persists in modified form. While modern technology provides advance warning of wildfires and tracks volcanic eruptions in real time, a blood moon still confirms that atmospheric particulate concentrations have reached levels affecting light transmission. This information remains relevant for health decisions, visibility concerns, and environmental awareness.

The next time you see the moon rising red, understand what you are witnessing. Somewhere, likely upwind from your location, fire burns. Forests combust, volcanoes erupt, or possibly cities flame. The fire has released countless particles into the atmosphere. These particles now float between you and the moon, filtering its reflected light, removing blue and green, leaving only red to reach your eyes. The ancients called it an omen and they were right. Fire has marked the atmosphere. The blood moon announces what the wind already knows: something burns. The sky bears witness, painting the moon crimson with the ash of conflagration. That ancient warning remains as valid now as ever. When the moon bleeds, fire speaks.

Chapter Eight: The Missing Dew

Farmers walking their fields at dawn have relied on dew presence for millennia. The cool dampness coating grass blades, leaves, and spider webs signals normal atmospheric moisture cycling. When those same farmers wake to completely dry vegetation despite clear skies and calm winds, they recognize danger. The missing dew announces that atmospheric humidity has dropped to critical levels, that soil moisture is depleting rapidly, and that flash drought threatens crops. This simple observation, requiring no instruments beyond human senses and attentive observation, provided early warning of agricultural disaster across countless generations. Modern meteorology confirms the physical processes underlying this traditional knowledge: dew formation depends on specific humidity and temperature relationships that break down when conditions favor rapid drought development.

Dew forms through condensation when objects cool below the dew point temperature of surrounding air. The dew point represents the temperature at which air becomes saturated with water vapor and condensation begins. During clear, calm nights, surfaces exposed to the sky radiate heat into space, cooling below air temperature. If surface temperature drops below the dew point, water vapor in contact with the surface condenses into liquid droplets, creating visible dew.

The process requires several conditions to occur. First, skies must be clear or nearly clear. Clouds block outgoing thermal radiation and prevent sufficient surface cooling. Second, winds must remain light. Strong winds mix air layers, preventing near-surface cooling. Third, sufficient atmospheric humidity must be present. Even if surfaces cool dramatically, if the air contains little water vapor, no dew forms. This third requirement links dew formation directly to atmospheric moisture content.

The dew point temperature provides direct measurement of atmospheric water vapor concentration. High dew points indicate abundant moisture. Low dew points signal dry air. When air temperature and dew point converge, relative humidity reaches 100 percent and condensation occurs. The wider the gap between air temperature and dew point, the lower the relative humidity and the less likely condensation becomes.

Flash drought develops when persistent high pressure systems stall over regions during growing season. These systems bring clear skies, light winds, and intense solar radiation. Daytime heating becomes extreme while nighttime cooling remains substantial. Soil moisture depletes rapidly through both evaporation and plant transpiration. Without precipitation to replenish soil water, the moisture deficit grows daily.

As soil dries, less water evaporates into the near-surface atmosphere. Plants begin conserving water by closing stomata, reducing transpiration. The combined effect drops atmospheric humidity substantially. Dew points fall as moisture content decreases. Eventually, even during optimal radiative cooling on clear, calm nights, surface temperatures cannot drop below the now very low dew point. Dew fails to form despite ideal meteorological conditions otherwise.

The absence of dew under clear, calm conditions therefore signals critically low atmospheric humidity. This low humidity results from depleted soil moisture and stressed vegetation, both indicators of developing drought. The progression from normal dew formation to intermittent dew to no dew tracks the intensification of moisture deficit. Farmers observing this sequence recognized escalating drought risk.

Historical agricultural societies developed acute sensitivity to dew patterns. Mediterranean farmers in ancient Greece and Rome noted dew presence as part of daily field inspections. Chinese agricultural texts spanning thousands of years reference dew observation as a farming fundamental. European peasants through the medieval and early modern periods included dew assessment in their morning routines. The knowledge appeared so basic that written instructions were unnecessary; it passed orally as common knowledge.

The correlation between missing dew and crop failure proved reliable enough that some cultures incorporated dew observation into religious practice. Morning prayers might include supplications for dew, recognizing its absence as harbinger of famine. Ritual attention to dew elevated the observation from simple meteorological monitoring to spiritual practice, ensuring that communities maintained vigilance.

Native American agricultural societies in North America developed similar practices. Tribes cultivating the Three Sisters corn, beans, and squash noted dew patterns throughout the growing season. Prolonged dew absence prompted adjustments in planting density, crop selection, and water management strategies. These observations integrated into broader environmental monitoring systems that tracked multiple indicators simultaneously.

The physics of radiative cooling that produces dew follows from fundamental thermodynamics. All objects emit thermal radiation proportional to their temperature. During daytime, solar radiation input exceeds outgoing thermal radiation, causing warming. At night, with no solar input, thermal radiation continues while energy input ceases, producing cooling. Clear skies allow maximum radiative cooling because greenhouse gases and clouds absorb and re-emit thermal radiation back toward the surface, reducing net cooling.

Water vapor acts as a significant greenhouse gas. Air with high humidity absorbs more outgoing thermal radiation than dry air, re-emitting some energy back to the surface. This effect reduces nighttime cooling. Conversely, very dry air allows more thermal radiation to escape to space, enhancing surface cooling. The irony is that dry air promotes stronger radiative cooling, but the lack of moisture means no dew forms despite colder surface temperatures.

The dew point temperature depression, the difference between air temperature and dew point, indicates how far temperature must drop for saturation to occur. On humid nights, the depression might be only a few degrees. Surface cooling of five degrees suffices for dew formation. On dry nights during drought, the depression can exceed thirty degrees. Even extreme radiative cooling cannot produce the temperature drop necessary for condensation.

Frost formation follows similar principles but occurs when surface temperatures drop below freezing. Water vapor condenses directly into ice crystals rather than liquid droplets, a process called deposition. The missing frost on clear, calm nights during freezing weather signals the same low humidity that produces missing dew during warmer months. The drought indicator functions across temperature ranges.

Agricultural impact from flash drought can be devastating. Crops experience water stress within days as soil moisture depletes. Yield losses accumulate rapidly, particularly for shallow rooted plants. Deep rooted perennials may access subsoil moisture temporarily but eventually suffer as the drought deepens. Without irrigation, farmers can only watch as crops wither.

The speed of flash drought development distinguishes it from conventional drought. Traditional drought develops over weeks to months as precipitation deficits accumulate gradually. Flash drought can establish in one to two weeks under intense heat and dry conditions. The rapidity leaves little time for adaptation. The missing dew provides one of the earliest readily observable warnings that flash drought has begun.

Modern drought monitoring uses complex indices incorporating precipitation, temperature, soil moisture, and vegetation health. The Palmer Drought Severity Index, Standardized Precipitation Index, and other metrics quantify drought conditions objectively. Satellite remote sensing reveals vegetation stress and soil moisture status. These technological tools provide comprehensive drought assessment impossible in historical periods.

Yet the simple observation of dew presence or absence remains valuable. It requires no technology, no specialized training, no infrastructure. Any farmer walking fields at dawn can assess dew conditions. The observation provides immediate, local information about current moisture status. While it cannot replace comprehensive monitoring, it supplements and validates technological assessments.

The timing of dew observation matters. Dew forms primarily in the hours before dawn when radiative cooling reaches maximum effect. Checking fields too early may not reveal whether dew will eventually form. Checking too late risks finding dew that has already evaporated after sunrise. The optimal observation window spans from roughly one hour before sunrise through the first hour after, balancing dew formation time against evaporation onset.

Geographic and topographic factors influence local dew patterns. Valleys accumulate cold air drainage, enhancing cooling and dew formation. Hilltops experience better air mixing, reducing dew. Areas near water bodies benefit from higher local humidity. Urban heat islands suppress dew through elevated temperatures. Farmers learned their specific fields' characteristics, understanding which locations formed dew most readily and which required more moisture to produce visible condensation.

Vegetation type affects dew formation and observation. Broad leaves collect more visible dew than narrow grass blades. Waxy leaf surfaces shed water more readily than rough, absorbent surfaces. Spider webs capture dew dramatically, making small amounts visible. Farmers knew which plants served as the best dew indicators for their region.

The distinction between dew and guttation sometimes caused confusion. Guttation occurs when plants exude water through specialized structures, typically appearing as droplets at leaf tips and edges. This happens when soil moisture is abundant and root pressure forces water out of plants. Guttation indicates high soil moisture, the opposite of drought conditions. Experienced observers distinguished between dew covering entire surfaces and guttation droplets at specific points.

Climate change affects dew patterns through multiple pathways. Rising temperatures increase atmospheric water holding capacity but also increase evaporative demand. Regional humidity patterns shift as precipitation regimes change. Some areas may see increased dew as moisture availability rises. Others may experience decreased dew as aridification proceeds. The changing baseline complicates interpretation of traditional signals.

Heat waves increasingly trigger flash drought events. The combination of extreme temperatures and stagnant high pressure creates conditions for rapid moisture depletion. Missing dew during heat waves provides confirmation that the situation has moved from uncomfortable heat into agricultural crisis. The sign retains its warning function even as the frequency of flash drought increases.

Urban development reduces areas where dew observation remains practical. Pavement, buildings, and landscaping alter surface temperature and moisture dynamics. City dwellers lose daily contact with dew patterns. This disconnect severs one more link to environmental literacy that historically informed human decision making.

The recovery of dew formation signals improving moisture conditions. When adequate precipitation falls, soil moisture replenishes, humidity rises, and dew resumes forming under appropriate meteorological conditions. Farmers observing the return of normal dew patterns after drought knew that crisis had passed and normal conditions were reestablishing.

The amount of dew provides additional information beyond simple presence or absence. Heavy dew soaking vegetation indicates high humidity and abundant moisture. Light dew barely visible suggests marginal conditions where moisture remains adequate but not abundant. Intermediate states provide gradations of information more nuanced than binary presence or absence assessments.

Scientific measurement of dew involves weighing absorbent materials before and after exposure, measuring moisture content directly. Meteorological stations sometimes include dew sensors that detect condensation automatically. These instruments quantify what farmers assessed qualitatively, translating observation into numerical data. Both approaches access the same underlying physical phenomenon, differing only in precision and documentation.

The next time you walk outside at dawn, notice whether dew covers vegetation, pavements, and objects. If conditions are clear and calm but surfaces remain completely dry, recognize what that absence signals. The atmosphere holds little moisture. Humidity has dropped to critical levels. Soil has dried. Drought threatens or has already begun. This simple observation, requiring only attention and awareness, provides information as valid today as when farmers first noticed the pattern thousands of years ago. The missing dew speaks clearly to those who know its language. The message remains unchanged: water grows scarce. The land thirsts. Without rain, withering follows.

Chapter Nine: The Contrail Clock

White lines streak across blue sky, marking the passage of aircraft at high altitude. These condensation trails, or contrails, form when hot, moist exhaust from jet engines meets frigid upper atmosphere air. Most observers dismiss contrails as mere artifacts of modern aviation, visual pollution marring otherwise pristine skies. Yet these artificial clouds carry information about atmospheric conditions at flight altitudes, information that correlates predictably with approaching weather systems. The behavior of contrails, specifically their persistence and spread, provides reliable indication of upper atmosphere humidity and can predict weather front arrival within twelve to thirty-six hours. Pilots and meteorologists understand this relationship, but the knowledge remains largely unknown to general populations who see contrails daily without recognizing their predictive value.

Contrails form through a combination of condensation and ice crystal nucleation. Jet engines burn hydrocarbon fuel, producing carbon dioxide and water vapor among other combustion products. The exhaust emerges at temperatures exceeding 500 degrees Celsius into ambient air at cruise altitude where temperatures typically range from negative forty to negative sixty degrees Celsius. This extreme temperature differential causes immediate condensation and freezing of water vapor.

The process mirrors breath condensation on cold days. Your warm, moist breath enters frigid air and instantly condenses into visible fog. Aircraft exhaust does the same but at far greater temperature extremes and with more water vapor. Additionally, exhaust contains particles that serve as condensation and ice nuclei, promoting droplet and crystal formation. The result is an artificial cloud extending behind the aircraft, marking its path across the sky.

Whether contrails form at all depends on specific atmospheric conditions at flight altitude. The air must be cold enough and humid enough for the added water vapor from exhaust to exceed saturation levels. This threshold varies with temperature and pressure but generally requires temperatures below negative forty degrees Celsius and relative humidity above roughly sixty percent with respect to ice. Below these thresholds, exhaust water vapor remains as invisible gas and no contrail forms.

The distinction between short lived and persistent contrails carries meteorological significance. Short lived contrails form but dissipate within minutes as ice crystals sublimate back to vapor. These indicate that the upper atmosphere, while cold enough for initial condensation, remains relatively dry. The ambient air cannot maintain ice crystals; they evaporate as exhaust diffuses into surrounding air. Short lived contrails suggest that high altitude air remains unsaturated, conditions typical of stable high pressure systems.

Persistent contrails behave entirely differently. They form and remain visible for hours, sometimes spreading to cover substantial portions of the sky. These contrails indicate that upper atmosphere air is near saturation or supersaturated with respect to ice. The ice crystals forming from exhaust do not sublimate because the ambient air already holds nearly maximum water vapor for the temperature. Instead, the crystals may actually grow as they encounter this humid air, exactly like natural cirrus clouds.

Persistent contrails that spread and expand signal extremely high upper atmosphere humidity, conditions virtually identical to those supporting natural cirrus cloud formation. This makes sense because cirrus clouds consist of ice crystals suspended in cold, humid air at high altitude. Persistent spreading contrails are essentially artificial cirrus clouds, following the same physics and responding to the same atmospheric conditions.

The meteorological significance emerges from understanding what upper atmosphere humidity predicts. High altitude moisture typically arrives ahead of weather fronts, particularly warm fronts. As frontal systems approach, moisture advances at multiple atmospheric levels. High altitude moisture appears first, often twelve to forty-eight hours before surface weather changes. This moisture allows persistent contrail formation where only hours earlier, short lived or no contrails appeared.

The progression from no contrails to short lived contrails to persistent contrails to spreading contrails maps directly onto increasing upper atmosphere humidity as weather systems approach. Observing this sequence provides advance warning that atmospheric conditions are evolving toward precipitation. The contrails serve as visible markers of invisible moisture increases at altitudes unreachable by surface observations.

Weather forecasters use water vapor satellite imagery to track upper atmosphere moisture. These satellite sensors detect infrared radiation emitted by water vapor molecules, creating images showing moisture distribution. Forecasters analyze these images to predict weather system movement and precipitation timing. Contrail behavior provides ground based observers access to similar information without requiring satellite data or specialized equipment.

The reliability of contrail based weather prediction depends on geographic location and prevailing weather patterns. The technique works best in temperate latitudes where frontal systems dominate weather. Regions with frequent high pressure dominance may see persistent contrails less often. Tropical areas with different weather mechanisms show different contrail patterns. However, anywhere that contrails form, their behavior still reflects current upper atmosphere conditions.

Military applications of contrail knowledge emerged during World War Two. Bomber crews flying at high altitude needed to understand when their aircraft would produce visible contrails that enemy fighters could use to locate them. Meteorologists developed contrail forecasts predicting altitudes and conditions where contrails would form. These forecasts required detailed understanding of temperature and humidity profiles through the atmosphere.

Post-war commercial aviation adopted similar considerations. Pilots sometimes adjust altitude to avoid persistent contrail formation when operating in sensitive areas or when fuel efficiency concerns arise. Contrail formation slightly increases drag and fuel consumption, though the effect is small. More significantly, persistent contrails contribute to radiative forcing, trapping outgoing thermal radiation and contributing to warming. Climate scientists study contrail impacts as part of aviation's overall climate footprint.

The conspiracy theory culture surrounding contrails demonstrates profound scientific illiteracy. "Chemtrail" believers claim that persistent contrails represent deliberate chemical spraying by governments or other organizations for sinister purposes. These claims ignore basic atmospheric physics and misinterpret natural contrail behavior as evidence of malicious intent. Debunking these theories requires explaining the legitimate science of contrail formation and persistence.

The reality is that contrail persistence variations result entirely from atmospheric humidity differences. The exact same aircraft flying the same route at the same altitude can produce short lived contrails one day and persistent spreading contrails the next, purely due to atmospheric moisture changes. No change in the aircraft, fuel, or operation occurs, only ambient conditions vary.

Contrail shadows provide additional atmospheric information. When the sun angle is appropriate, contrails cast shadows onto lower cloud layers or onto the ground. These shadows can appear as dark parallel lines moving across the landscape. The visibility of contrail shadows indicates specific geometric relationships between sun position, contrail altitude, and observer location. While less directly useful for weather prediction than contrail persistence, shadows confirm contrail altitude and thickness.

The color of contrails varies with lighting conditions and viewing angle. Backlit contrails appear darker, sometimes nearly black against bright sky. Illuminated contrails appear brilliant white. Contrails catching sunrise or sunset light glow orange or pink, creating dramatic visual effects. These color variations result from light scattering through ice crystals and provide no meteorological information beyond confirming ice crystal presence.

Multiple contrails at different altitudes sometimes appear simultaneously, revealing atmospheric stratification. Lower contrails may dissipate quickly while higher contrails persist, or vice versa, indicating humidity variations with altitude. Sharp contrail boundaries show distinct air mass differences. Fuzzy, spreading contrails reveal turbulent mixing. These observations provide three dimensional information about atmospheric structure.

The spacing between individual contrail lines from four engine aircraft reveals atmospheric stability. In calm conditions, the four trails remain distinct and parallel. In turbulent conditions, trails blur together quickly. Wind shear stretches contrails into curves or distorts their shapes. Strong upper level winds carry contrails downwind rapidly, sometimes creating herringbone patterns across the sky.

Distrails, or dissipation trails, form when aircraft descend through cloud layers. The aircraft passage introduces dry air or mixing that causes existing clouds to evaporate along the flight path, creating a clear slot through the cloud deck. Distrails indicate clouds exist at marginally stable conditions where small perturbations cause dissipation. This suggests that the weather system producing those clouds may be weakening or moving away.

The width of contrails depends on aircraft type, engine specifications, and atmospheric conditions. Larger aircraft with bigger engines produce wider contrails. High humidity allows contrails to expand more readily. Initial contrail width within seconds of formation reveals exhaust characteristics. Subsequent spreading over minutes to hours reveals atmospheric moisture and mixing.

Night contrails occur but remain largely invisible unless illuminated by moonlight or artificial lighting. The ice crystals exist but without illumination, human eyes cannot detect them. Infrared sensors can detect night contrails through their thermal signature. Military aircraft operating at night benefit from contrail invisibility, avoiding visual detection that daylight contrails provide to observers.

Contrail cirrus, the term for extensive artificial cloud cover from persistent spreading contrails, can cover significant fractions of the sky over busy air routes. This artificial cloudiness affects local climate through both solar reflection and thermal absorption. Contrail cirrus is most extensive over heavily trafficked air routes in regions with frequent upper atmosphere humidity. The east coast United States and western Europe see particularly dense contrail cirrus development.

Climate models incorporate contrail cirrus effects when assessing aviation impacts. The net effect appears to be warming because contrails block more outgoing thermal radiation than incoming solar radiation. However, the magnitude and regional distribution of these effects remain active research topics. Understanding contrail formation and behavior becomes increasingly important as air traffic grows globally.

Educational efforts around contrails aim to combat misinformation while promoting atmospheric literacy. Teaching populations to observe contrail behavior as an indicator of upper atmosphere conditions serves multiple purposes. It builds scientific understanding, provides practical weather forecasting skills, and counters conspiracy narratives through knowledge of actual physical processes.

The next time you notice aircraft contrails, watch their behavior over the following minutes and hours. Do they dissipate quickly, vanishing within a minute or two? This indicates relatively dry upper atmosphere, conditions favoring continued fair weather. Do they persist and gradually spread, eventually covering portions of the sky? This signals high upper atmosphere humidity, warning that weather systems approach within the next day or two. The contrails function as a clock, marking time until atmospheric conditions change. Like so many atmospheric omens, they speak clearly to observers who understand their language. The jets write temporary messages across the sky. Those messages describe invisible moisture beginning its journey downward, moisture that will eventually reach the surface as rain or snow. The contrail clock ticks toward weather change. Learn to read its hands.

Chapter Ten: Aurora Whispers

The Northern Lights shimmer across polar skies in curtains of green, red, and purple, mesmerizing observers with their silent dance. Across the globe, the Southern Lights mirror this display over Antarctic regions. These auroras result from charged particles from the sun colliding with Earth's upper atmosphere, exciting atoms that then emit light. The visual spectacle has inspired mythology, art, and scientific investigation for centuries. But some observers report something beyond the visual display: they claim to hear the auroras, describing crackling sounds, hisses, and whispers synchronized with visible movements. Scientists long dismissed these reports as impossible, arguing that auroras occur at altitudes far too high for sound to propagate to ground observers. Yet accumulating evidence suggests certain individuals do hear something real, a phenomenon called electrophonic hearing that challenges conventional understanding of sound and perception.

Auroras form when solar wind particles, primarily electrons and protons, interact with Earth's magnetic field and atmosphere. The sun continuously emits charged particles in all directions as solar wind. When directed at Earth, these particles encounter our planet's magnetic field, which deflects most around the planet. However, magnetic field lines converge near the poles, funneling particles into the upper atmosphere at high latitudes.

These charged particles collide with atmospheric atoms and molecules at altitudes between 100 and 400 kilometers. The collisions transfer energy to electrons in oxygen and nitrogen atoms, raising them to excited states. When these electrons return to lower energy states, they emit photons of specific wavelengths determined by atomic structure. Oxygen produces green and red light. Nitrogen produces blue, purple, and occasionally red. The specific colors depend on collision energy and atmospheric altitude.

The visual display varies from faint glows to brilliant curtains rippling across half the sky. Auroras can remain relatively static or change rapidly, forming rays, arcs, and coronas. During intense geomagnetic storms, auroras become visible at lower latitudes, sometimes reaching as far south as the Mediterranean or northern United States. The intensity correlates with solar activity, particularly coronal mass ejections that send massive bursts of charged particles toward Earth.

Sound propagation from aurora altitudes to the ground faces severe physical constraints. Sound requires a medium to travel through, propagating as pressure waves through air molecules. At 100 kilometers altitude where most auroras occur, air density is approximately one millionth that at sea level. This extremely thin atmosphere cannot support sound waves with sufficient energy to reach the ground. Additionally, the temperature profile of the atmosphere affects sound propagation, and the distances involved would attenuate any sound far beyond detectability.

These facts led scientists to confidently dismiss aurora sound reports for decades. Witnesses describing crackling or hissing sounds were assumed to be experiencing auditory hallucinations, psychological effects of witnessing awe inspiring visual phenomena, or misattributing unrelated environmental sounds to the auroras. The physics seemed conclusive: auroras were too high and too quiet to be heard.

However, reports persisted across cultures and centuries. Inuit populations in Arctic regions consistently described aurora sounds. Norse sagas reference whispers accompanying the lights. Siberian indigenous peoples incorporated aurora sounds into cultural narratives. These reports shared common characteristics: the sounds were faint, localized, and synchronized with rapid aurora movements. They were described as crackling, clapping, rustling, or hissing. Multiple observers sometimes heard them simultaneously while others nearby heard nothing.

The geographic distribution of reports provided a clue. Observers in dry, cold environments during periods of low humidity reported sounds more frequently than those in humid conditions. Mountain observers reported sounds more often than those at sea level. These patterns suggested environmental factors affecting whether sounds could be detected rather than the auroras themselves producing audible sounds in conventional ways.

Research in the 1970s and 1980s began exploring alternative explanations. Some scientists proposed that electromagnetic fields associated with auroras might induce electrical currents in metallic objects like eyeglasses or jewelry, causing vibrations that produced faint sounds. Others suggested that very low frequency electromagnetic waves could potentially stimulate auditory nerves directly, creating the perception of sound without actual acoustic waves.

The electrophonic hearing hypothesis gained support from observations of similar phenomena associated with other electromagnetic events. Witnesses to meteor fireballs occasionally report hearing hissing or crackling sounds simultaneous with the visual display, despite meteors being dozens of kilometers distant. These sounds cannot be ordinary acoustic propagation given the distances and timing. The leading explanation invokes electromagnetic waves generated by the meteor interacting with objects near the observer to create localized sounds.

Finnish researcher Unto Laine conducted systematic investigations beginning in the 1990s. He placed microphones in locations where observers frequently reported aurora sounds. The microphones occasionally recorded faint crackling sounds during periods of intense aurora activity and rapid movement. These recordings provided the first objective documentation of sounds correlated with auroras, though their mechanism remained unclear.

Laine proposed that temperature inversions near the ground could trap electromagnetic effects in the lowest atmosphere. During clear, cold nights typical of aurora viewing, temperature inversions frequently develop with warmer air above colder surface air. This layered structure might affect how electromagnetic fields associated with auroras interact with the surface environment. Charged particles separated in temperature inversion layers could discharge suddenly as aurora activity fluctuates, creating acoustic waves directly at ground level.

This mechanism requires specific atmospheric conditions: strong temperature inversions, very low humidity, calm winds, and of course, active auroras overhead. These conditions co-occur infrequently, explaining why aurora sounds are rarely reported even in aurora prone regions. Most aurora observers never hear anything. Those who do tend to be in optimal locations during unusual atmospheric conditions.

The sounds themselves provide little audible information beyond confirming electromagnetic activity. They do not encode data about solar wind properties or geomagnetic field variations in any accessible way. However, their presence confirms that the aurora overhead is particularly active, that geomagnetic conditions are disturbed, and by extension, that solar activity is elevated. This third order information had limited practical utility for historical populations.

However, severe geomagnetic storms can disrupt technological systems. Power grids experience induced currents that can damage transformers. Communication systems suffer interference. Satellite operations encounter difficulties. Modern infrastructure vulnerable to space weather effects faces genuine risks during major geomagnetic storms. The aurora sounds, when they occur, signal that such a storm is underway.

The Carrington Event of 1859 represents the most severe geomagnetic storm in recorded history. Telegraph systems worldwide experienced disruptions, some operators receiving electric shocks, some equipment catching fire. Auroras appeared at tropical latitudes. Contemporary accounts mention unusual sounds and sensations during the event. While difficult to verify given 19th century scientific limitations, these reports may represent genuine electrophonic phenomena during an extraordinarily intense geomagnetic storm.

If a Carrington level event occurred today, technological infrastructure would face catastrophic damage. Power grids could fail across continents. Communication networks would suffer extensive disruption. Satellites might be permanently disabled. The economic impact would total trillions of dollars. While such extreme events remain rare, occurring perhaps once per century, their consequences have grown as society has become more technologically dependent.

The omens read from aurora sounds would thus be warnings of electromagnetic disturbance sufficient to affect technology. Pre-industrial societies had little vulnerable infrastructure, so the warning carried minimal practical value. Modern civilization, in contrast, depends critically on systems vulnerable to geomagnetic effects. The ancient omen has gained new relevance.

Cultural interpretations of aurora sounds varied widely. Some traditions treated them as voices of spirits or ancestors. Others considered them warnings or messages from deities. Norse mythology incorporated auroras as reflections from Valkyrie armor, and sounds might represent their battle cries. These supernatural frameworks provided meaning to observations that had no scientific explanation within those cultures' knowledge systems.

Inuit traditions treated auroras with caution and respect. Some groups believed the lights could be summoned by whistling but warned against attracting their attention. Others taught that auroras represented the spirits of the dead and should not be disturbed. These beliefs encoded behavioral responses, discouraging activities that might prove dangerous in the extreme cold of aurora viewing conditions, though expressed through spiritual rather than safety rationales.

The psychological impact of experiencing both sight and sound from auroras amplifies their impressiveness. Witnessing a powerful aurora is already awe inspiring. Adding auditory elements creates multisensory immersion that deepens emotional impact. This enhanced experience likely reinforced cultural significance wherever aurora sounds were detected, ensuring that reports persisted across generations.

Modern aurora chasers, photographers and tourists who travel to high latitudes specifically to observe Northern Lights, occasionally report sound experiences. Online forums contain accounts from multiple independent observers describing similar phenomena. While individual reports remain anecdotal, the aggregate pattern supports that something genuine occasionally occurs, even if imperfectly understood.

Research into electrophonic phenomena continues, hampered by the difficulty of studying rare, unpredictable events. Recording equipment must be deployed in remote locations for extended periods, hoping to capture the right combination of atmospheric and geomagnetic conditions. The sounds themselves, when they occur, are faint and brief, easily masked by environmental noise. Controlled experiments proving the mechanism remain challenging to design and execute.

The next solar maximum, the phase of the sun's eleven year activity cycle when sunspots and solar eruptions peak, will provide increased opportunities to investigate aurora phenomena including potential sounds. Space weather monitoring has improved dramatically in recent decades, allowing better prediction of when auroras will be visible and potentially audible. Coordinated observation campaigns may finally resolve remaining questions about how and why some observers hear the lights.

The philosophical implications of electrophonic hearing extend beyond auroras. If electromagnetic fields can create perceived sounds through direct neural stimulation or through interaction with nearby objects, what other sensory experiences might result from electromagnetic exposure? Do some people possess greater sensitivity to such effects? Could electromagnetic fields affect cognition or mood through mechanisms we don't yet understand? These questions touch fundamental issues about perception, consciousness, and environmental interactions.

The aurora omen, then, speaks to those rare individuals in rare circumstances through whispers and crackles synchronized with celestial lights. The message is simple: the electromagnetic environment has become disturbed. Solar particles cascade through Earth's magnetic field. The upper atmosphere glows. And occasionally, under just the right conditions, that disturbance manifests as faint sounds at ground level. For most of human history, the warning had no practical significance. Today, it signals potential threats to technological civilization. The whispers of the aurora now carry weight they never held before. When the Northern Lights speak, perhaps we should finally learn to listen.

Chapter Eleven: Dust Devils

Small whirlwinds dance across open ground, spinning columns of dust and debris skyward in tight vortices. These dust devils appear suddenly on hot afternoons, moving erratically across fields, parking lots, and deserts before dissipating as quickly as they formed. Most observers regard them as minor curiosities, harmless atmospheric oddities providing momentary entertainment. Yet dust devils carry information about surface conditions, wind patterns, and atmospheric instability. Their frequency, intensity, and behavior signal soil erosion vulnerability, wind shear potential, and thermal structure of the lowest atmosphere. Agricultural societies historically recognized dust devil patterns as indicators of land condition and weather trends. The swirling omens revealed which fields had degraded, where winds would strengthen, and when atmospheric instability threatened more severe weather.

Dust devils form through intense surface heating combined with wind shear near the ground. Solar radiation heats exposed soil or pavement to temperatures significantly above ambient air. This creates a shallow layer of superheated air in contact with the hot surface. When this hot air becomes buoyant enough, it rises rapidly, creating an updraft. If even slight rotation exists in the wind field, conservation of angular momentum concentrates this rotation as air converges toward the updraft, exactly like water spinning faster as it approaches a drain.

The resulting vortex stretches vertically, becoming tighter and more intense. The low pressure at the vortex core draws in surrounding air near the surface while the updraft carries air upward. This circulation lifts dust, sand, and loose debris, making the invisible vortex visible. Without sufficient loose material to entrain, dust devils remain transparent, detectable only through their effects on vegetation or through distortion of background images by density variations.

Dust devil formation requires several specific conditions. First, strong surface heating must create substantial temperature differences between the ground and air just above it. Second, winds must be light but not calm, providing the initial rotation that intensifies during vortex formation. Third, the surface must be relatively uniform over distances of tens to hundreds of meters, allowing unobstructed heat buildup. Fourth, atmospheric stability above the surface layer must permit vertical motion without excessive resistance.

The intensity of dust devils varies enormously. Small vortices barely a meter across last seconds and lift only the finest dust particles. Large dust devils can exceed one hundred meters in height, span ten or more meters in diameter, and persist for twenty minutes or longer. The strongest dust devils generate winds exceeding sixty miles per hour at the core, sufficient to damage structures or overturn vehicles. These intense vortices approach weak tornadoes in strength, though they form through different mechanisms.

The distinction between dust devils and tornadoes is fundamental. Tornadoes form from rotating thunderstorms, descending from cloud base to ground. They require substantial atmospheric instability, moisture, and specific wind shear profiles. Tornadoes are associated with severe weather systems and carry enormous destructive potential. Dust devils form in fair weather from surface heating alone. They build from the ground upward rather than descending from clouds. While they can be locally intense, they lack the organized storm structure that produces the most violent tornadoes.

The seasonal and diurnal patterns of dust devil activity reveal their dependence on surface heating. They occur most frequently during late spring, summer, and early fall when solar radiation is most intense. Within each day, peak activity occurs in early to mid afternoon when surface temperatures reach maximum values. Mornings and evenings see few dust devils as surfaces cool.

Cloudy days suppress dust devil formation by reducing surface heating.

Geographic distribution follows patterns of strong surface heating and appropriate wind conditions. Deserts produce frequent dust devils because exposed soil heats intensely and vegetation doesn't shade the surface. Plowed agricultural fields generate dust devils when soil is dry and winds are light. Urban areas with extensive pavement experience dust devils in parking lots and along roadways. Any surface that can heat strongly and lacks obstacles to vortex development can support dust devil formation.

The correlation between dust devil frequency and soil condition provides the basis for the agricultural omen. Healthy soil with good structure, adequate organic matter, and vegetation cover resists erosion. Even when dust devils form over such soil, they lift minimal material. Degraded soil, lacking structure and exposed without protective cover, provides abundant fine particles that dust devils readily entrain. Fields showing frequent, dust laden devils signal erosion vulnerability.

Farmers observing their fields could assess soil condition through dust devil characteristics. Dark, dense dust clouds indicated fine soil particles becoming airborne, evidence of soil structure breakdown. Pale, sparse dust suggested better soil condition with less erodible material available. The persistence of dust devils over particular areas identified spots where soil had degraded most severely, guiding remediation efforts.

Dust devils also reveal surface wind patterns invisible otherwise. The vortices move with wind flow at low altitudes, tracking air movement near the ground. Observing multiple dust devils across a landscape shows how wind channels through terrain features, where flow converges or diverges, and where shear zones exist. This information helped predict where winds would strengthen and where crops or structures faced greatest wind exposure.

The presence of strong dust devils indicates atmospheric conditions that can support other convective phenomena. The same intense surface heating and instability that generates vigorous dust devils can, under appropriate moisture conditions, produce thunderstorms. While dust devils themselves form in fair weather, their intensity reveals how much thermal energy the lower atmosphere contains. This energy, if combined with moisture and lifting mechanisms, could fuel severe convection.

Native American groups inhabiting arid regions incorporated dust devils into cultural frameworks. Navajo tradition associates dust devils with spirits and teaches specific responses when encountering them. These cultural practices encoded behavioral responses appropriate for desert environments, using spiritual narratives to transmit practical knowledge about atmospheric conditions and land stewardship.

Dust devils on Mars have been extensively photographed by rovers and orbiters. The Martian atmosphere, though much thinner than Earth's, still supports vortex formation through surface heating and wind shear. Martian dust devils can tower many kilometers high and leave distinctive tracks across the surface. Studying these extraterrestrial vortices helps planetary scientists understand Martian atmospheric dynamics and surface processes. The comparison with terrestrial dust devils reveals both similarities in fundamental physics and differences arising from atmospheric composition and pressure.

The ability to forecast dust devil activity has limited practical utility for most applications. They cause minor disruptions occasionally, perhaps overturning lightweight objects or briefly reducing visibility. However, for some operations, dust devil prediction matters. Airport operations near desert regions must account for sudden visibility reductions and wind gusts. Outdoor events in open areas face potential disruption. Agricultural operations involving spraying or loose materials need to avoid peak dust devil times.

Climate change affects dust devil occurrence through multiple pathways. Rising temperatures increase surface heating intensity, potentially enhancing dust devil formation. Changing precipitation patterns alter soil moisture and vegetation cover, affecting both surface temperatures and erodible material availability. Desertification in some regions creates more areas conducive to dust devil development. The net effect remains an active research question.

Dust devils transport significant quantities of soil and dust into the atmosphere. While individually small compared to dust storms, their cumulative effect over large areas and long timescales contributes meaningfully to atmospheric dust loading. This dust affects climate through radiative effects, serves as ice nuclei in clouds, transports nutrients across ecosystems, and impacts air quality. Understanding dust devil contributions to these processes requires quantifying their frequency, size distribution, and dust lifting capacity.

Measurement and documentation of dust devils pose challenges. Their unpredictable formation and brief lifespans make systematic observation difficult. Automated detection systems using cameras or meteorological sensors can capture events humans would miss, building statistical databases of dust devil properties. These data reveal patterns in occurrence rates, size distributions, and temporal variations that inform theoretical understanding.

The record sizes for observed dust devils remain uncertain because measurements are rarely available during the largest events. Anecdotal reports describe vortices exceeding two hundred meters in height and twenty meters in diameter, with winds approaching tornado strength. Confirming such reports requires instrumented encounters that rarely occur. The distinction between large dust devils and weak tornadoes becomes blurred at the extreme end of the spectrum.

Photography and video documentation of dust devils has exploded with smartphone technology. Citizens now capture and share thousands of dust devil images and videos, providing unprecedented documentation of these phenomena. While most lack scientific measurements, the visual records reveal geographic patterns, size ranges, and behavioral characteristics that supplement formal studies.

The sound of dust devils varies with size and intensity. Small vortices produce gentle rushing sounds as air circulates. Larger, more intense devils generate roaring sounds from rapid wind motion and debris impact. Observers caught inside dust devils report distinctive acoustic experiences, with wind noise surrounding them and the relatively calm vortex core creating acoustic contrast. These experiences, while interesting, carry no predictive information beyond confirming the dust devil's presence.

Educational opportunities around dust devils remain underutilized. These accessible atmospheric phenomena occur frequently enough that students can observe them during appropriate seasons. Understanding their formation combines thermodynamics, fluid mechanics, and observational meteorology in a tangible example. Science education could leverage dust devil observation to teach fundamental atmospheric concepts through direct experience rather than abstract theory.

The next time you see a dust devil spinning across open ground, recognize what it reveals. The surface has heated intensely, creating strong temperature gradients in the lowest atmosphere. Winds at ground level contain rotation that concentrates into visible vortices. The soil or surface material is sufficiently loose and dry to be readily entrained into the circulation. These conditions tell stories about the land beneath your feet, about moisture deficits, about wind patterns, about atmospheric energy. Ancient farmers read these signs to assess their fields and predict when winds would strengthen or storms might develop.

That knowledge remains as valid today as millennia ago. The dust devil is a small omen, localized and brief. But it speaks truthfully about conditions where it forms. Those who understand its language gain insight that supplements and validates technological monitoring. The vortex traces a temporary message in the air, written in dust and wind. The message is simple: the land here is dry, the sun here is strong, the air here is unstable. Heed it.

Chapter Twelve: Solar Storms

The sun appears constant in the daily sky, rising and setting with reliable predictability. Yet our star is anything but stable. Its surface roils with magnetic activity, occasionally unleashing massive eruptions that hurl billions of tons of charged particles into space. When these eruptions direct toward Earth, they trigger geomagnetic storms that can disable satellites, disrupt power grids, interfere with navigation systems, and produce auroras at unexpected latitudes. Historical records document how solar storms affected telegraph systems in the 19th century, demonstrating that space weather poses genuine threats to technological infrastructure. Modern civilization's dependence on electricity, communications, and satellite technology creates unprecedented vulnerability to severe solar storms. The ancient omens of unusual auroras and magnetic disturbances now signal potential catastrophe for interconnected global systems.

Solar storms originate from the sun's magnetic field dynamics. The sun consists of plasma, ionized gas where electrons separate from atomic nuclei. This plasma conducts electricity and responds to magnetic fields. The sun's rotation, which varies with latitude, twists and tangles magnetic field lines threading through the solar interior and atmosphere. These twisted fields store enormous energy, sometimes releasing it catastrophically through solar flares and coronal mass ejections.

Solar flares represent sudden releases of magnetic energy as reconfigured field lines snap into lower energy states. The energy heats plasma to tens of millions of degrees and accelerates particles to near light speed. Flares emit radiation across the electromagnetic spectrum, from radio waves through visible light to X-rays and gamma rays. The most intense flares can be seen as bright spots on the solar surface when viewed through appropriate filters. Flare radiation reaches Earth in approximately eight minutes, traveling at light speed.

Coronal mass ejections, or CMEs, are more relevant for geomagnetic storm effects than flares themselves. CMEs involve massive eruptions of plasma and magnetic field from the sun's corona. These eruptions can contain billions of tons of material traveling at speeds from several hundred to over two thousand kilometers per second. If a CME directs toward Earth, the ejected material takes one to four days to traverse the sun-Earth distance of approximately 150 million kilometers.

When a CME reaches Earth, it encounters our planet's magnetic field. The interaction depends on the CME's magnetic field orientation. If the CME field aligns opposite to Earth's field, magnetic reconnection occurs, allowing solar plasma to penetrate the magnetosphere more effectively. This drives the geomagnetic storm, compressing Earth's magnetic field on the sunward side and stretching it into a long tail on the night side. The disturbance injects energy into the magnetosphere, which eventually dissipates through various processes including aurora formation and induced electrical currents.

These induced currents, called geomagnetically induced currents or GICs, flow through any conductive pathways including power transmission lines, pipelines, and telecommunications cables. GICs appear as quasi-DC currents superimposed on the alternating current normally flowing in power systems. Transformers in power grids are particularly vulnerable because they're designed for AC operation and can saturate when DC currents flow through them. Saturation produces excessive heating, potentially damaging or destroying transformers that take months or years to replace.

The Carrington Event of September 1859 remains the most powerful geomagnetic storm in recorded history. British astronomer Richard Carrington observed a massive solar flare on September 1. Seventeen hours later, effects reached Earth, indicating an extremely fast CME. Telegraph systems worldwide experienced disruptions. Operators reported receiving electric shocks. Some telegraph equipment caught fire. Telegraph messages could be sent even with power supplies disconnected, the induced currents providing sufficient voltage.

Auroras appeared at tropical latitudes during the Carrington Event. Observers in Cuba, Hawaii, and Colombia reported seeing Northern Lights, phenomena normally confined to high latitudes. The auroras were so bright that newspapers could be read by their light at night. Birds began chirping, confused by the illumination. Miners woke and prepared breakfast, thinking dawn had arrived.

If a Carrington level event occurred today, consequences would be catastrophic. Power grids across North America, Europe, and Asia could fail simultaneously as GICs damaged transformers faster than operators could respond. The cascading failures would leave hundreds of millions without electricity for weeks or months while replacement transformers were manufactured and installed. Economic losses would exceed two trillion dollars, potentially triggering global recession.

Satellite systems face direct threats from solar storms. Charged particles damage solar panels and electronic components through cumulative radiation exposure and single event upsets where individual cosmic rays flip bits in computer memory. Atmospheric heating during geomagnetic storms increases drag on low Earth orbit satellites, accelerating orbital decay. Communication disruptions occur as ionospheric disturbances affect radio wave propagation. GPS accuracy degrades as positioning calculations rely on precise timing signals affected by ionospheric conditions.

The March 1989 geomagnetic storm demonstrated vulnerability on a smaller scale than Carrington but still significant. The storm knocked out power to Quebec for nine hours, leaving six million people without electricity. Transformers at the Salem Nuclear Power Plant in New Jersey were damaged. Satellites experienced malfunctions. Auroras appeared as far south as Texas and Florida. The event provided clear warning that modern infrastructure faces genuine risk from space weather.

Airlines operate with awareness of space weather because solar storms affect flight operations. Polar routes expose crews and passengers to increased radiation during solar storms, sometimes exceeding regulatory limits. Airlines reroute flights to lower latitudes during severe events, accepting longer flight times and increased fuel costs to ensure safety. Communication disruptions affect air traffic control, particularly over polar regions and oceans where satellite links provide primary communications.

Historical records preserve evidence of solar storm effects before the technological era. Ice cores and tree rings contain isotopes produced when solar particles interact with the atmosphere. These proxies reveal that events potentially larger than Carrington have occurred. A solar storm in 774 CE left distinctive signatures in multiple independent records. Another in 993 CE shows similar evidence. These events occurred when human infrastructure was invulnerable to space weather, but they demonstrate that the sun can produce storms exceeding anything witnessed in the modern era.

The sun's activity follows an eleven year cycle, alternating between solar minimum with few sunspots and solar maximum with abundant sunspots and increased flare and CME activity. However, individual solar cycles vary dramatically in intensity. Some maxima produce numerous strong events while others remain relatively quiet. Predicting cycle strength in advance remains challenging despite decades of research. The current solar cycle, Cycle 25, began in 2019 and will peak around 2025, though exact timing and intensity remain uncertain.

Space weather forecasting has improved substantially through dedicated spacecraft monitoring the sun. The Solar and Heliospheric Observatory, or SOHO, observes the sun continuously from a gravitationally stable point between Earth and sun. The Solar Dynamics Observatory provides high resolution images revealing active regions likely to produce flares. When CMEs occur, the Advanced Composition Explorer satellite at the same gravitationally stable point provides approximately one hour advance warning before the CME reaches Earth by detecting the approaching disturbance.

This warning time, while better than nothing, remains insufficient for comprehensive protective measures. Power grid operators can shed load and reconfigure systems to reduce vulnerability, but major storms can overwhelm defenses faster than human operators respond. Satellites can enter safe modes, but radiation damage accumulates regardless. The fundamental problem is that technological systems remain vulnerable to effects we can predict only imperfectly and mitigate only partially.

Mitigation strategies exist but require substantial investment. Power grids can install additional transformers as spares, allowing rapid replacement of damaged units. Neutral current blocking devices can prevent GIC flow. System designs can incorporate resilience through redundancy and rapid reconfiguration capability. However, implementing these measures across entire interconnected grids costs billions of dollars, funds that utilities and governments often deem better spent elsewhere until disaster strikes.

The geopolitical implications of severe solar storms deserve consideration. A Carrington level event would affect multiple nations simultaneously, straining international cooperation during crisis response. Nations with more resilient infrastructure or better preparation would recover faster, potentially shifting economic and political balances. The event would be truly global in scope, offering no safe havens, a unique characteristic among natural disasters.

Cultural responses to historical geomagnetic storms varied based on scientific understanding. The Carrington Event occurred before anyone understood space weather or even accepted that the sun was a nearby star rather than something fundamentally different. Witnesses interpreted the auroras through available frameworks, seeing spiritual significance or supernatural manifestations. Today's scientific understanding strips away mystery but reveals genuine threats that superstitious interpretations missed.

The next major solar storm will serve as test of modern preparedness and resilience. Decades of warnings and planning will prove either adequate or insufficient. The probability of a Carrington level event remains low in any given year, perhaps one percent, but over decades the cumulative probability becomes substantial. We are not prepared for an inevitability we know will eventually occur but cannot predict precisely.

The omen interpretation of auroras as warnings gains new meaning in technological context. Historical observers seeing auroras at unusual latitudes correctly identified abnormal conditions but lacked vulnerable infrastructure for those conditions to damage. Modern observers seeing similar auroras should recognize them as warnings that space weather threatens technological systems. The sign remains the same. The consequences have changed dramatically.

Citizen science projects now recruit volunteers to observe and report auroras, contributing to space weather monitoring. These observations supplement instrumental data and engage public interest in phenomena that directly affect modern life. Education about space weather builds awareness that can drive policy changes and infrastructure investments before catastrophe forces reactive measures.

The next time you witness aurora displays at latitudes where they rarely appear, understand what you are seeing. The sun has hurled plasma and magnetic fields toward Earth. Those fields now interact with our planet's magnetic environment, channeling energy into our atmosphere and inducing currents in any conducting paths. Technology throughout the hemisphere faces stress. If the aurora is brilliant and widespread, if it appears in colors and forms rarely witnessed, the storm is severe. That ancient omen, the lights in the sky that inspired fear and wonder in our ancestors, now carries practical warning that modern civilization should heed. The sun has not changed its behavior. We have changed our vulnerability. The omen speaks as loudly as ever. We have simply given it new things to threaten. When the Northern Lights dance at unexpected latitudes, when the sky glows with colors that do not belong there, recognize the message. The sun has reached out to touch our technological world. And its touch, however beautiful, carries danger we are only beginning to understand.

Chapter Thirteen: Atmospheric Rivers

Satellite imagery reveals them as vast plumes of moisture streaming across thousands of miles of ocean, ribbons of concentrated water vapor flowing through the atmosphere like celestial rivers. These atmospheric rivers transport enormous quantities of moisture from tropical regions toward mid latitudes, carrying as much water as the Amazon River at any given moment. When these rivers make landfall, they can produce torrential rainfall, catastrophic flooding, and landslides that reshape terrain. California experiences atmospheric rivers regularly, receiving between thirty and fifty percent of its annual precipitation from just a few such events. Yet the same systems that bring needed water can overwhelm flood control infrastructure and cause billions of dollars in damage. Historical populations living in atmospheric river prone regions recognized patterns in storm sequences, cloud formations, and wind directions that signaled when megafloods threatened. These omens of the atmospheric river announced when normal storms would intensify into disasters of biblical proportions.

Atmospheric rivers form when moisture laden air from tropical or subtropical oceans organizes into narrow corridors thousands of kilometers long but only a few hundred kilometers wide. The mechanisms creating these structures involve complex interactions between mid latitude weather systems and tropical moisture sources. Extratropical cyclones, the low pressure systems that dominate mid latitude weather, extend frontal boundaries that can tap tropical moisture reservoirs. When positioned correctly, these systems act as pumps, drawing moisture poleward in concentrated streams.

The water vapor content in atmospheric rivers exceeds typical atmospheric moisture concentrations by factors of two to five. The narrow geometry and high moisture loading mean these features transport water with remarkable efficiency. A strong atmospheric river might carry water flux equivalent to ten to twenty Mississippi Rivers, though of course the water travels as vapor in the atmosphere rather than liquid in channels. This enormous transport capacity allows atmospheric rivers to deliver massive precipitation totals when they encounter mountains or other lifting mechanisms that force the air upward and wring out the moisture.

The western coast of North America provides ideal conditions for atmospheric river impacts. Prevailing westerly winds carry Pacific moisture toward the continent. The coastal mountain ranges force incoming air upward, cooling it and triggering condensation and precipitation. When an atmospheric river makes landfall perpendicular to the coast and remains stationary for hours or days, the combination of continuous moisture supply and orographic lifting produces rainfall totals that can exceed twenty inches in forty-eight hours. Such amounts overwhelm drainage systems and trigger widespread flooding.

The Sierra Nevada mountains experience particularly intense atmospheric river impacts. The range rises steeply from the Central Valley to elevations exceeding 14,000 feet. Atmospheric rivers bringing moisture to these mountains produce prodigious snowfall at high elevations and heavy rain at lower elevations. The snow provides essential water storage for California's dry summers, but excessive rainfall at lower elevations causes floods while rapid warming after atmospheric river events can trigger catastrophic snowmelt flooding.

Historical megafloods in California resulted from extreme atmospheric river events. The Great Flood of 1862 inundated California's Central Valley, creating an inland sea up to twenty miles wide and three hundred miles long. Atmospheric river after atmospheric river battered the region for forty-five days, delivering rainfall totals exceeding one hundred twenty inches in some mountain locations. The state capital moved temporarily from flooded Sacramento. Economic losses, adjusted for inflation, would exceed ten billion dollars today.

Native American oral traditions in California preserve knowledge of great floods occurring at irregular intervals. These traditions describe waters covering vast areas, people fleeing to high ground, and landscapes transformed by inundation. The traditions encode practical knowledge about flood risk and appropriate settlement locations, transmitted across generations through narrative frameworks that ensured preservation and attention.

Geological evidence confirms that megafloods occurred repeatedly throughout California's pre-historical record. Sediment deposits, erosion patterns, and lake bed stratigraphy reveal flood events far exceeding anything witnessed since European settlement. These paleofloods occurred during different climate regimes but demonstrate that the physical processes enabling catastrophic flooding have operated for millennia.

Forecasting atmospheric river impacts has improved through better understanding of their structure and behavior. Meteorologists now recognize that not all atmospheric rivers are equal. Some remain narrow, move quickly, and produce moderate precipitation. Others are wider, stronger, and more persistent, producing the extreme precipitation that causes disasters. A classification system ranks atmospheric rivers from AR1, weak, through AR5, exceptional, based on moisture transport and duration. AR4 and AR5 events pose the greatest flood risks.

The challenge is that atmospheric rivers provide both essential water resources and flood hazards. California's water supply depends on atmospheric river precipitation to fill reservoirs and build mountain snowpack. Too few atmospheric rivers means drought. Too many, or a few extreme ones, means flooding. Managing this balance requires sophisticated forecasting to predict event intensity and duration, allowing reservoir operators to adjust water storage levels proactively.

The pineapple express represents a specific type of atmospheric river originating near Hawaii and directed toward the Pacific Northwest and British Columbia. These events derive their name from moisture source near the Hawaiian Islands. Pineapple express events bring warm, moist air that can produce rain rather than snow even at higher elevations, eliminating the moderating effect of snowfall and maximizing flood potential. The January 1997 pineapple express caused widespread flooding throughout the Pacific Northwest, costing billions in damages.

European coasts also experience atmospheric river impacts, though they receive less attention than Pacific events. The United Kingdom and Ireland encounter atmospheric rivers originating over the Atlantic. These systems bring heavy rain and strong winds, causing flooding in river basins and coastal areas. The winter storms that batter Britain often involve atmospheric river moisture, though they may not be explicitly identified as such in weather reporting.

The frequency and intensity of atmospheric rivers may change as climate warms. Warmer atmosphere holds more water vapor, following the Clausius-Clapeyron relationship where saturation vapor pressure increases exponentially with temperature. This enhanced moisture availability could intensify atmospheric rivers, producing higher peak precipitation rates. However, changes in circulation patterns might also alter atmospheric river tracks and frequency, making future projections uncertain.

Research using paleoclimate proxies attempts to understand how atmospheric rivers behaved under different climate conditions. Speleothems, stalactites and stalagmites from caves, record precipitation patterns through oxygen isotope ratios in their layered growth. Tree ring patterns reveal wet and dry periods spanning centuries. Sediment cores from lakes and ocean floors preserve flood deposits. These records suggest that atmospheric river intensity and frequency varied substantially across different climate epochs.

The omens that warned of megafloods involved multiple observable phenomena. Unusually persistent southwesterly winds over several days indicated atmospheric river approach. Clouds taking on specific characteristics, particularly long bands of mid level clouds oriented perpendicular to mountains, signaled moisture transport. Rapid temperature increases during winter storms suggested warm subtropical air arrival, meaning more rain instead of snow and higher flood potential. Experienced observers noting these patterns understood that standard storm preparations would prove inadequate.

Modern warning systems for atmospheric rivers combine satellite observations, weather modeling, and impact forecasting. Satellites equipped with microwave sensors detect atmospheric moisture content over oceans, identifying developing atmospheric rivers days before landfall. Numerical weather prediction models simulate their evolution and predict where they will strike coastlines. Hydrological models estimate resulting river flows and flood potential. These technological tools provide warning times measured in days rather than hours.

However, uncertainty remains substantial. Small shifts in atmospheric river track can mean the difference between moderate useful rain and catastrophic flooding. Changes in atmospheric river intensity as it approaches land are difficult to predict accurately. The duration of atmospheric river conditions, critical for determining total precipitation, often surprises forecasters. These uncertainties complicate emergency management decisions about evacuations, flood control operations, and resource positioning.

Urban development in flood prone areas has increased vulnerability to atmospheric river flooding. Cities expand into floodplains, paving surfaces and channeling runoff into streams unable to accommodate extreme flows.

Infrastructure ages beyond design life, reducing its effectiveness during extreme events. As populations grow in atmospheric river prone regions, the potential for disasters increases even if event frequency remains constant.

Reservoir management during atmospheric rivers requires sophisticated decision making. Operators must balance flood control, water supply, environmental flows, and hydroelectric generation objectives simultaneously. As an atmospheric river approaches, operators face decisions about releasing water to create flood storage capacity versus holding water to maximize supply during the subsequent dry season. Wrong decisions can either fail to prevent flooding or waste precious water that will be needed later.

The ecological impacts of atmospheric rivers extend beyond flooding. The moisture supports ecosystems adapted to Mediterranean climate patterns of wet winters and dry summers. Trees, particularly coastal redwoods, capture fog and drizzle associated with atmospheric river systems.

Salmon runs depend on flow pulses from atmospheric river precipitation to trigger migration and provide access to spawning grounds. The ecological relationships woven into atmospheric river patterns developed over thousands of years.

Climate adaptation strategies must account for changing atmospheric river behavior. Infrastructure designed for historical flood magnitudes may prove inadequate if atmospheric rivers intensify. Reservoir operating rules based on past precipitation patterns may not suit future conditions. Building codes and flood plain management regulations require updating to reflect evolving risks. These adaptations demand resources and political will that often materialize only after disasters demonstrate need.

The next time forecasters predict an atmospheric river making landfall, understand what approaches. A river of vapor thousands of miles long converges on the coast. It carries moisture drawn from warm tropical oceans, pumped poleward by weather systems that will soon arrive. When that moisture encounters mountains, it will condense and fall as rain or snow measured in feet rather than inches. If the conditions align incorrectly, if the river remains stationary too long, if the moisture content exceeds expectations, what begins as needed precipitation transforms into catastrophic flooding. The ancient omens pointed toward this danger through observable signs in clouds, winds, and temperatures. Modern technology provides earlier and more precise warnings, but the fundamental message remains unchanged. An atmospheric river approaches. It carries both blessing and curse. Water comes, in quantities that can refresh or destroy. Preparation determines which outcome prevails. The megaflood omen still speaks. We would be wise to listen.

Chapter Fourteen: The Smell of Ozone

A sharp, clean scent fills the air during thunderstorms, distinctive and unmistakable to anyone who has experienced it. This smell, often described as metallic or like bleach, signals that lightning strikes nearby. The odor results from ozone formation when electrical discharges split oxygen molecules and recombine them into the three atom form that characterizes this reactive gas. For thousands of years, humans have associated this smell with approaching electrical storms, recognizing it as warning to seek shelter immediately. The olfactory omen provided advance notice measured in seconds to minutes before lightning struck close enough to pose lethal danger. Modern understanding confirms that ozone concentration spikes dramatically during thunderstorms, validating the ancient knowledge that the electric smell announces deadly peril overhead.

Ozone consists of three oxygen atoms bonded together, forming the molecular formula O3.

Normal atmospheric oxygen exists as two atom molecules, O2. The additional oxygen atom makes ozone chemically reactive and unstable. It readily releases the extra atom, reverting to stable O2 while oxidizing whatever substance it encounters. This reactivity makes ozone both useful as a disinfectant and dangerous as a respiratory irritant at high concentrations.

Lightning generates ozone through multiple mechanisms. The primary process involves the electrical discharge itself. Lightning channels carry currents exceeding 20,000 amperes and heat air to temperatures around 30,000 Kelvin, five times hotter than the sun's surface. These extreme conditions split oxygen and nitrogen molecules into constituent atoms. Some oxygen atoms recombine with O2 molecules to form O3. Additionally, ultraviolet radiation emitted by the lightning discharge splits oxygen molecules, and the freed atoms form ozone through subsequent reactions.

The distinctive smell of ozone becomes detectable at concentrations as low as 0.01 to 0.02 parts per million. Human olfactory sensitivity to ozone exceeds sensitivity to most other atmospheric gases, an evolutionary feature that may have provided survival advantage by warning of electrical storm danger. Lightning struck objects and nearby ground strikes create localized ozone concentrations far exceeding detection thresholds, sometimes reaching several parts per million.

Downbursts and downdrafts from thunderstorms carry ozone generated at cloud level down to ground level. Thunderstorm updrafts lift air to altitudes where lightning occurs most frequently. The ozone formed there descends with precipitation and downdrafts, bringing the smell to observers on the ground even before lightning strikes near them. This mechanism allows the smell to arrive minutes before nearby strikes, providing valuable warning time.

The timing and intensity of ozone smell provide information about storm proximity and intensity. Faint ozone detected during clear conditions might indicate distant storms with ozone transported by winds. Sudden strong ozone smell during active storms signals nearby lightning activity. The strongest ozone odors occur immediately after very close strikes, within hundreds of meters, as freshly formed ozone spreads from the discharge channel.

Historical accounts of thunderstorms frequently mention the distinctive smell, though observers lacked chemical knowledge to identify it as ozone. Greek and Roman natural philosophers described electrical storm odors without understanding their source. Medieval chroniclers noted the smell in storm descriptions. Sailors, shepherds, and farmers, those who spent extensive time outdoors exposed to weather, developed acute sensitivity to the smell and recognized its warning value.

The chemical identification of ozone occurred in the 1840s. German chemist Christian Friedrich Schönbein investigated the smell produced by electrical sparks and identified a new substance he named ozone from the Greek word meaning "to smell." He demonstrated that ozone formed through electrical discharge in oxygen and documented its chemical properties. This work provided scientific foundation for understanding the storm smell that humans had noticed for millennia.

Ozone in the lower atmosphere, called tropospheric ozone, differs fundamentally from stratospheric ozone despite being chemically identical. Stratospheric ozone, located ten to thirty miles above Earth's surface, forms through ultraviolet radiation from the sun and protects life from harmful UV exposure. Tropospheric ozone at ground level acts as a pollutant, damaging lungs and plant tissues. Lightning represents a natural source of tropospheric ozone, though human activities now dominate ozone production near populated areas through photochemical reactions involving vehicle emissions and industrial pollutants.

The concentration of ozone produced by lightning varies with storm intensity. Severe storms with high flash rates generate more ozone than weaker storms. Supercells with continuous lightning activity for hours can produce ozone plumes detectable downwind for tens of kilometers. The cumulative effect of numerous lightning flashes in a small area creates ozone hotspots that persist after storms pass, slowly dissipating through chemical reactions and dispersion.

Thunder provides additional warning of lightning danger, but the ozone smell can precede thunder for nearby strikes. Sound travels approximately one kilometer every three seconds in standard conditions. The delay between lightning and thunder indicates distance: a three second delay means roughly one kilometer distance. Ozone, transported by wind, can arrive before sound from the same discharge, particularly if wind blows from the lightning toward the observer. Thus the smell might provide slightly earlier warning than sound for some strikes.

The ozone warning operates on very short timescales compared to other atmospheric omens. While many signs provide hours or days of advance notice, ozone smell indicates immediate danger. The appropriate response is to seek substantial shelter instantly. No time exists for gradual preparation or monitoring of developing conditions. The smell announces that lethal electrical discharges are occurring close enough that the next strike could hit the observer's location.

Lightning safety procedures emphasize the 30-30 rule: seek shelter when the time between lightning flash and thunder is thirty seconds or less, stay sheltered for thirty minutes after the last thunder. The ozone smell provides complementary information, confirming lightning proximity when visibility is limited or thunder is masked by other sounds. Sheltering when strong ozone smell appears, even without seeing lightning or hearing thunder, represents appropriate caution.

Enclosed buildings with plumbing and electrical wiring provide the best lightning protection. The conductive paths through the structure allow lightning current to reach ground without passing through occupants. Vehicles with metal roofs also provide good protection through similar principles. Open structures, trees, and isolated objects in open terrain offer no protection and should be avoided. The time to reach appropriate shelter after detecting ozone smell may be insufficient if observers are far from buildings when storms approach.

The physiological effects of ozone at concentrations produced by thunderstorms generally remain below levels causing acute symptoms. Ozone irritates respiratory passages, triggering coughing and throat irritation at concentrations above 0.1 parts per million. Very high concentrations can cause chest pain and breathing difficulty. However, the ozone levels produced by lightning typically cause only minor irritation if any, especially considering the brief exposure duration. The smell serves as a warning rather than a direct hazard in most circumstances.

Ozone chemistry in the atmosphere involves complex reaction chains. Ozone molecules react with many atmospheric constituents including nitrogen oxides, volatile organic compounds, and water vapor. These reactions remove ozone while producing other compounds, some of which contribute to air quality problems. The ozone produced by lightning undergoes these same reactions, typically dissipating within hours to days depending on atmospheric conditions and available reactants.

Measurements of ozone concentration during thunderstorms confirm substantial increases. Research aircraft flying near storms detect ozone levels several times higher than background concentrations. Ground stations downwind from intense storms record ozone spikes correlating with lightning activity. These measurements validate the connection between electrical discharge, ozone formation, and the storm smell that observers detect.

The global production of ozone by lightning contributes meaningfully to atmospheric ozone budgets. Estimates suggest lightning produces several million tons of ozone annually worldwide. This natural source competes with anthropogenic sources in regions with heavy pollution but dominates in remote areas far from human activities. Understanding lightning contributions to atmospheric chemistry requires accounting for this ozone production in models of air quality and atmospheric composition.

Climate change may alter lightning frequency through effects on atmospheric instability and convection. Warmer temperatures increase atmospheric moisture and energy available for thunderstorm development. Some climate models predict increased lightning activity as warming proceeds, though regional variations complicate global projections. Increased lightning would produce more ozone, potentially affecting tropospheric chemistry and air quality in ways that compound other climate related atmospheric changes.

The connection between ozone smell and lightning safety takes on renewed importance as populations increasingly engage in outdoor recreation. Hikers, campers, and athletes exercising outdoors face lightning risk, particularly in mountain regions where afternoon thunderstorms develop regularly. Educating these populations about ozone smell as a warning sign could reduce lightning casualties, which still claim dozens of lives annually in the United States alone.

Technology provides alternative lightning warning through smartphone apps, weather radios, and specialized lightning detection systems. These tools track lightning strikes in real time and issue warnings when strikes occur within specified distances. However, technology requires functional devices, cellular service, and user attention. The ozone smell requires only a working nose and awareness of its significance. As a backup warning system requiring no infrastructure, the olfactory omen retains value even in a technological era.

The next time you smell that sharp, clean odor during a storm, recognize it immediately for what it is. Lightning has split oxygen molecules nearby. Ozone has formed and now reaches your nose, carried by wind or downdraft. Electrical discharges intense enough to restructure atmospheric chemistry are occurring within striking distance. The next bolt might target the ground near you, seeking the path of least resistance to earth. No time exists for contemplation or gradual response. The smell is the warning. Seek shelter now. Move to an enclosed building or vehicle. Avoid open ground, isolated trees, and tall objects. Do not wait for visible lightning or audible thunder. The ozone has announced the danger. Historical populations who heeded this warning survived to pass the knowledge forward. Those who ignored it became cautionary tales. The electric smell speaks with urgent clarity. Its message never varies. Lightning strikes nearby. Shelter immediately. Your life depends on recognizing this ancient omen and responding with appropriate haste. The air itself warns you. Listen with your nose and move toward safety.

Chapter Fifteen: The Vanishing Horizon

The desert horizon dissolves into shimmering uncertainty, solid ground appearing to float above liquid surfaces that do not exist. Distant objects waver and distort, mountains invert, cities appear suspended in the sky. These mirages have misled travelers for millennia, promising water where none exists, distorting distances, and creating fantastical visions that challenged observers' grasp on reality. The most complex mirages, called Fata Morgana after the Arthurian sorceress Morgan le Fay, transform landscapes into impossible architectures of towers, castles, and cliffs where nothing stands. Yet mirages are not hallucinations or supernatural manifestations. They result from atmospheric refraction when temperature gradients bend light rays along curved paths. Understanding mirage formation and behavior provides essential survival information for desert travel, revealing temperature conditions, surface characteristics, and atmospheric stability. The vanishing horizon announced danger to those who could read its warnings: heat stress threatened, water remained distant, and navigation by visual landmarks would prove unreliable.

Mirages form when light travels through air layers with different temperatures and densities. Light moves at slightly different speeds through air of different densities, following from the basic physics of refraction. When light passes from one density to another at an angle, it bends at the boundary. In the atmosphere, where temperature and thus density change continuously with height, light follows curved rather than straight paths. The curvature becomes visible as mirages when temperature gradients are sufficiently steep.

Inferior mirages, the most common type, occur when ground surface heats intensely, creating a thin layer of hot, low density air near the surface with cooler, denser air above. Light from the sky traveling downward toward the ground bends upward as it encounters progressively less dense air. If the temperature gradient is steep enough, light rays curve strongly enough that they never reach the ground, instead bending back upward toward the observer. The observer sees sky light appearing to come from the ground surface, creating the appearance of water or a reflective surface.

The mechanism mirrors how light bends when entering water at an angle, except the density transition occurs gradually through a layer rather than abruptly at a surface. The hot air near the ground acts as a less dense medium, bending light away from the normal direction just as water bends light toward the normal. The continuous gradient produces continuous curving rather than the single angular bend at a sharp interface.

Desert travelers seeing inferior mirages observe what appears to be shimmering water on the road or ground ahead. As they approach, the apparent water retreats, maintaining distance. This behavior results from the viewing geometry: the mirage appears at a specific angle below horizontal determined by the temperature gradient. As the observer moves, the point where their line of sight at that angle intersects the ground moves correspondingly. The mirage cannot be reached because it is not an object at a fixed location but rather a viewing angle phenomenon that shifts with observer position.

The shimmer and wavering characteristic of inferior mirages result from turbulence in the hot air layer. Temperature doesn't remain uniform but fluctuates as hot and cool air parcels mix through convection. These fluctuations change the refractive index along light paths, causing the mirage to dance and waver. The effect resembles looking through moving water, which also creates wavering distortions through refractive index variations.

Superior mirages form under inverse temperature conditions where cold dense air sits below warmer less dense air. These conditions occur over cold water surfaces or snow covered ground with warmer air above. Light from distant objects curves downward as it travels through progressively denser air near the surface. This downward curving allows observers to see objects beyond the geometric horizon, as light follows a path that dips below a straight line and then rises to reach the observer.

Superior mirages make distant objects appear elevated, sometimes dramatically so. Ships beyond the horizon become visible, appearing to float in the sky. Coastlines distort vertically, with cliffs and buildings stretching upward. The effect can make small waves appear as towering walls of water. These mirages occur commonly in polar regions and over cold ocean surfaces, areas where surface temperatures remain significantly below air temperatures at higher altitudes.

Fata Morgana represents the most complex mirage type, involving multiple temperature inversions creating multiple curved light paths from the same object. The result is vertically stacked and distorted images that can transform simple objects into elaborate structures. A distant ship might appear as multiple stacked images, some upright, some inverted, creating the appearance of an impossible vessel or floating castle. Rocky coastlines become towers and ramparts of an ethereal city.

The name Fata Morgana derives from Arthurian legend, with Morgan le Fay purportedly creating phantom castles through sorcery to lure sailors to their doom. The mirages frequently observed in the Strait of Messina between Italy and Sicily were attributed to Morgan's magic. These displays result from the strait's geography, with cold water flowing at depth and warmer air masses above, creating ideal conditions for complex mirage formation.

Arctic explorers encountered Fata Morgana that created dangerous navigational errors. Land that appeared close and accessible actually lay far beyond reach. Ice formations distorted beyond recognition. The 19th century search for the Northwest Passage saw multiple expeditions confused by mirages that made navigation by visual landmarks nearly impossible. Explorers recorded detailed descriptions of phantom mountains and coastlines that vanished as conditions changed.

The Crocker Land expedition of 1913 sought a land mass that Admiral Robert Peary claimed to have observed from northern Ellesmott Island in 1906. The expedition found nothing. Peary almost certainly observed a Fata Morgana, mistaking miraged ice formations or distant land for undiscovered territory. The incident demonstrates how compelling mirages appear even to experienced observers with scientific training.

Desert cultures worldwide developed sophisticated understanding of mirage behavior. Tuareg nomads crossing the Sahara distinguished between different mirage types and understood their limitations. Bedouin travelers in Arabian deserts recognized mirage conditions and adjusted navigation accordingly. Aboriginal Australians traversing arid interior regions incorporated mirage awareness into traditional knowledge systems. These populations survived in part through understanding atmospheric optical phenomena that could otherwise prove fatal through navigational errors.

The survival implications of mirages extend beyond simple navigation errors. Inferior mirages indicating extreme surface heating warn of dangerous heat conditions. Air temperatures at ground level might reach lethal levels during the hottest part of the day. The presence of strong mirages announces that continuing travel during those hours risks heat stroke. Traditional desert travel patterns involved moving during early morning and late evening, sheltering during midday heat. Mirage intensity provided visual confirmation of when surface conditions became dangerous.

The absence of mirages under expected conditions could signal unusual atmospheric states. Overcast skies preventing surface heating would suppress mirage formation. High humidity increasing air density might reduce mirages even during hot conditions. These variations from expected patterns provided information about atmospheric conditions that might affect weather development, water availability from dew or fog, and animal behavior.

Distance estimation becomes severely compromised during mirage conditions. Objects appear closer or farther than actual positions. Elevated mirages make distant landmarks appear nearer, encouraging travelers to believe destinations are within reach when many additional hours of travel remain. This misjudgment could prove fatal if water supplies were calculated based on apparent rather than actual distances.

Modern vehicles reduce but don't eliminate mirage related dangers. Highway mirages cause drivers to perceive water on pavement ahead, sometimes prompting inappropriate braking or swerving. Drivers unfamiliar with desert conditions might slow dramatically or stop, creating collision risks. Aircraft pilots must account for mirages during landing approaches in desert regions, as runway appearance distorts through multiple refractive effects.

Optical instruments including cameras and telescopes are affected by mirages similarly to human vision. Astronomical observations from ground based telescopes degrade when temperature gradients near the telescope create turbulence and mirage effects. Site selection for major observatories considers temperature stability to minimize these distortions. Desert astronomy faces particular challenges during daytime observations when solar heating creates extreme temperature gradients.

The physics of mirage formation applies to other atmospheric phenomena. The green flash at sunset involves refraction through atmospheric layers. Halos around the sun and moon result from refraction through ice crystals. All these effects share the common principle that light bends when traveling through media with varying refractive indices. Understanding mirages therefore illuminates broader principles of atmospheric optics.

Climate change affects mirage occurrence through impacts on surface temperatures and atmospheric stability. Increasing extreme heat events intensify surface heating, potentially making mirages more frequent and pronounced in already arid regions. Changing precipitation patterns alter soil moisture and surface characteristics, affecting how surfaces heat and cool. Desertification expanding into previously marginal areas brings mirage conditions to regions where populations lack traditional knowledge for interpreting them.

Photography and video documentation of mirages sometimes fails to capture what observers see directly. Cameras and eyes process light differently, with human vision including psychological interpretation that cameras lack. Complex mirages might appear less dramatic in photographs than they did to observers experiencing them directly. Conversely, camera sensors sometimes record details invisible to human eyes, revealing structure in miraged images that observers cannot distinguish.

The next time you encounter a vanishing horizon, when the boundary between earth and sky loses definition and certainty, recognize what you witness. Extreme temperature gradients bend light along curved paths. The atmosphere refracts like a gigantic lens, distorting reality and creating false images. What appears solid may be illusion. What seems near may be impossibly distant. The mirage announces environmental conditions through visual distortion. Strong inferior mirages declare intense surface heating and dangerous thermal stress. Complex Fata Morgana warn that atmospheric layering prevents reliable navigation by sight. The ancients who survived desert crossings learned these lessons through experience, often bitter. They recognized that when the horizon vanished, when the world transformed into wavering uncertainty, trusting eyes alone invited disaster. Shelter from heat, wait for stable conditions, navigate by sun and stars rather than distant landmarks. The mirage speaks truth through falsehood, warning precisely by deceiving. Those who understand this paradox, who see through the illusion to the physical reality creating it, gain survival advantage. The vanishing horizon marks the boundary between those who comprehend atmospheric omens and those who wander confused through a world that no longer makes visual sense. Cross that boundary with knowledge, or perish in beautiful, shimmering ignorance.

Bibliography

Ahrens, C. Donald. Meteorology Today: An Introduction to Weather, Climate, and the Environment. Cengage Learning, 2012.

Burroughs, William J. Climate Change in Prehistory: The End of the Reign of Chaos. Cambridge University Press, 2005.

Dutton, John A. The Ceaseless Wind: An Introduction to the Theory of Atmospheric Motion. Dover Publications, 1986.

Frisinger, H. Howard. The History of Meteorology to 1800. American Meteorological Society, 1977.

Hamblyn, Richard. The Invention of Clouds: How an Amateur Meteorologist Forged the Language of the Skies. Picador, 2001.

Lutgens, Frederick K., and Edward J. Tarbuck. The Atmosphere: An Introduction to Meteorology. Pearson, 2015.

McIlveen, Robin. Fundamentals of Weather and Climate. Oxford University Press, 2010.

Monmonier, Mark. Air Apparent: How Meteorologists Learned to Map, Predict, and Dramatize Weather. University of Chicago Press, 1999.

Moran, Joseph M. Weather Studies: Introduction to Atmospheric Science. American Meteorological Society, 2017.

National Research Council. Severe Space Weather Events: Understanding Societal and Economic Impacts. National Academies Press, 2008.

Pretor-Pinney, Gavin. The Cloudspotter's Guide: The Science, History, and Culture of Clouds. Perigee Books, 2006.

Robinson, Andrew. Earth Shock: Climate, Complexity and the Forces of Nature. Thames & Hudson, 1993.

Suplee, Curt. Everyday Science Explained. National Geographic Society, 1996.

Williams, Jack. The AMS Weather Book: The Ultimate Guide to America's Weather. American Meteorological Society, 2009.

About the Author

Shane Frye is a writer and researcher specializing in the study of omens, atmospheric phenomena, and the intersection of traditional knowledge with modern science. His work explores how ancient cultures interpreted natural signs and how those interpretations often contained genuine observational wisdom that modern meteorology and atmospheric science have confirmed.

With a focus on making complex scientific concepts accessible while honoring historical perspectives, Frye examines how humanity has always looked to the sky for warnings, guidance, and understanding. His research spans multiple disciplines including meteorology, history, anthropology, and cultural studies, revealing the sophisticated environmental literacy that pre-scientific societies developed through careful observation.

The Sky is Falling: Omens of the Atmosphere represents Frye's comprehensive examination of fifteen atmospheric phenomena that have shaped human history and influenced civilizations. Through detailed analysis of both the physical mechanisms behind these signs and their cultural interpretations across diverse societies, Frye demonstrates that atmospheric omens remain as relevant today as they were millennia ago.

Frye continues his work documenting how natural signs speak to those who understand their language, bridging the gap between ancient wisdom and contemporary knowledge.

www.ingramcontent.com/pod-product-compliance
Lightning Source LLC
LaVergne TN
LVHW100308210926
848480LV00003B/1059